Design and Application of Combined Catalysis

Design and Application of Combined Catalysis

Guest Editors

Feng Wang
Xiyu Cheng
Jianhua Hu

MDPI

Basel • Beijing • Wuhan • Barcelona • Belgrade • Novi Sad • Cluj • Manchester

Guest Editors

Feng Wang
School of Food and Biological Engineering
Jiangsu University
Zhenjiang
China

Xiyu Cheng
College of Life Sciences and Bioengineering, School of Physical Science and Engineering
Beijing Jiaotong University
Beijing
China

Jianhua Hu
School of Chemical Engineering
Inner Mongolia University of Technology
Hohhot
China

Editorial Office
MDPI AG
Grosspeteranlage 5
4052 Basel, Switzerland

This is a reprint of the Special Issue, published open access by the journal *Catalysts* (ISSN 2073-4344), freely accessible at: https://www.mdpi.com/journal/catalysts/special_issues/T8G715NWM6.

For citation purposes, cite each article independently as indicated on the article page online and as indicated below:

Lastname, A.A.; Lastname, B.B. Article Title. *Journal Name* **Year**, *Volume Number*, Page Range.

ISBN 978-3-7258-6912-1 (Hbk)
ISBN 978-3-7258-6913-8 (PDF)
https://doi.org/10.3390/books978-3-7258-6913-8

Contents

catalysts

MDPI

Editorial

Editorial for the Special Issue "Design and Application of Combined Catalysis"

Feng Wang

School of Food and Biological Engineering, Jiangsu University, Zhenjiang 212013, China; fengwang@ujs.edu.cn

The global community is currently confronting the dual challenges of energy shortages and environmental degradation, underscoring an urgent need for green and sustainable technological solutions [1,2]. To break away from dependence on fossil resources and build a resilient energy system, green catalytic processes are regarded as a transformative pathway [3]. These processes can meet energy demands while simultaneously alleviating environmental pressures [4]. Green catalytic technologies have been widely applied in the chemical industry, pharmaceuticals, the energy sector, and environmental protection. However, conventional single-catalytic modes—such as photocatalytic, electrocatalytic, enzymatic, or chemocatalytic approaches—though promising, often suffer from inherent limitations, including low efficiency, poor stability, unsatisfactory selectivity, or high costs, which severely hinder their large-scale implementation [5,6].

To overcome these barriers, the field is increasingly turning to combinatorial catalysis. This innovative approach integrates distinct catalytic modalities, such as enzymatic, chemical, photocatalytic, electrocatalytic, and others, to create synergistic effects that are unattainable by any single system alone [7,8]. By uniting, for example, photocatalysis with electrocatalysis, enzymatic with heterogeneous catalysis, or whole-cell systems with synthetic catalysts, it becomes possible to orchestrate reaction networks that leverage the unique advantages of each component [9,10]. These strategies not only enhance reaction activity, selectivity, and durability but also create novel reaction pathways. Furthermore, integrated catalysis contributes to reduced energy consumption and minimized waste generation, aligning perfectly with the core objectives of sustainable chemistry [11].

This Special Issue, titled "Design and Application of Combined Catalysis", highlights the expanding role of combinatorial catalysis in addressing the inefficiencies of single-catalytic modes and reducing catalysis costs across diverse fields. Bringing together advances in catalyst design (e.g., magnetic immobilization and doping modification), innovative synthesis strategies, and mechanistic insights into multi-catalysis synergy, this Special Issue exhibits the dynamic evolution of combinatorial catalysis systems. Beyond showcasing technical progress, it underscores the importance of combinatorial catalysis as a key tool for achieving efficient compound production through sustainable approaches. The research presented in this Special Issue vividly demonstrates how the synergistic interplay between different catalysts can lead to remarkable enhancements in activity, stability, and selectivity, thereby paving the way for more sustainable and economically viable chemical processes.

In Contribution 1, a systematic study of two glycosidase immobilization methods—cross-linked enzyme aggregates (CLEAs) and magnetic cross-linked enzyme aggregates (MCLEAs)—revealed that MCLEAs eliminate the need for ultrafiltration, enhance immobilization efficiency and catalytic activity (approximately 30% higher than CLEAs), and maintain over 70% activity after 10 reuse cycles, providing an efficient and stable biocatalytic strategy for icaritin production.

Contribution 2 employed ion beam sputtering as a post-synthesis surface modification technique to enhance the low-temperature activity of commercial Pt/CeO_2–ZrO_2 catalysts. Using low-energy Ar^+ ions, the authors selectively generated oxygen vacancies and improved Pt dispersion without altering the bulk structure. The treated catalysts showed a progressive reduction in light-off temperatures for CO and C_3H_6 oxidation, with the most extensively irradiated samples achieving 13 and 22 K decreases, respectively, highlighting the potential of ion beam processing to address the persistent challenge of cold-start emissions in high-efficiency diesel and lean-burn engines.

In Contribution 3, Aleithan et al. synthesized MoS_2 nanohybrids via chemical vapor deposition and combined them with Au nanoparticles to construct MoS_2/Au hybrid materials for enhancing the performance of the electrocatalytic hydrogen evolution reaction (HER). They found that Au incorporation significantly enhances the conductivity and active sites of composites, enabling outstanding HER activity in neutral media. The onset potential reaches as low as −0.152 V with an exchange current density of 0.22 mA/cm^2, accompanied by excellent long-term stability. This work provides a viable strategy for developing efficient, stable non-precious metal-assisted electrocatalysts.

Contribution 4 investigated Mg-doped β-Bi_2O_3 nanoparticles prepared using sol–gel synthesis. The sol–gel-synthesized Mg-doped Bi_2O_3 samples showed a reduction in crystallite size from 79 to 13 nm and bandgap narrowing from 3.8 to 3.08–3.3 eV. The optimized $Mg_{0.075}Bi_{1.925}O_3$ catalyst achieved a high photodegradation rate constant of 0.0217 min^{-1} for methylene blue and exhibited enhanced hydrogen evolution due to the optimal band alignment structure. This study highlights the role of alkaline earth metal doping in designing efficient photocatalysts for environmental and energy applications.

Contribution 5 employed density functional theory (DFT) to explore the oxygen evolution reaction (OER) mechanism on NiP surfaces, revealing synergistic catalysis. P atoms stabilized O* intermediates, while Ni-induced electron donation enhanced the O*-P interaction. Moreover, Lewis acidic sites (from electron loss at adsorption sites) reduced overpotential by up to 0.61 V, with the Ni_4P_2 surface showing the lowest overpotential (0.78 V) via the single-site pathway. This study establishes a valuable framework for guiding base metal catalyst design.

Contribution 6 reported reduced graphene oxide (RGO)-coated ZrO_2 heterostructures synthesized by sol–gel combustion and ultrasonic mixing, where synergistic catalysis between RGO and ZrO_2 led to a reduced ZrO_2 bandgap from 5.6 to 3.87 eV and ZrO_2@4%RGO achieved 80% methylene blue (MB) degradation in 120 min under sunlight, while the reaction rate constant was nearly twice that of pure ZrO_2 (65%).

In Contribution 7, $TiO_2/(MA)_2SnCl_4$ nanocomposite films were synthesized via sunlight-driven drop-casting, where the synergistic effects of structural modification, optimized charge dynamics, and defect-mediated electronic interactions between TiO_2 and $(MA)_2SnCl_4$ led to a narrowed bandgap (from 3.1 to 2.6 eV) and enhanced visible-light absorption. The composite achieved 90% MB degradation in 60 min, outperforming pure TiO_2 (75%) due to Sn-O-Ti interactions and oxygen vacancies.

Contribution 8 demonstrated that Gd-doped ZnO nanocomposites, synthesized via sol–gel combustion, exhibit enhanced photocatalytic activity. Gd doping (0.025–0.075) modulated the bandgap (with a minimum of 3.07 eV for the 0.075 sample) and inhibited electron–hole recombination, with 0.075 Gd-doped ZnO degrading 89% MB in 120 min under UV light (40% for pure ZnO) and maintaining stability over five reuse cycles.

The final contribution (Contribution 9) established the optimal conditions for the laccase-catalyzed degradation of 2-MIB (pH 4.0, 25 °C, 150 rpm, 0.1 U/mL laccase, and 200 ng/L 2-MIB), achieving a degradation efficiency of approximately 78% within 4 h. The degradation efficiency was further elevated to 90.78% upon the application of a micro-electric field (0.04 A), demonstrating a notable synergistic effect between the enzymatic and electrocatalytic processes.

Taken together, the contributions in this Special Issue provide mechanistic insights, practical knowledge, and technological innovations that advance the field of combinatorial catalysis. They demonstrate how multi-catalysis synergy, material modification, and process optimization can improve catalytic efficiency and reduce costs.

I wish to extend my deepest and most sincere gratitude to all the authors for their invaluable contributions, without which this Special Issue would not have been possible. I hope that the original articles included in this issue will help advance the understanding of and resolve the current challenges in the field. I also sincerely thank MDPI's journal *Catalysts* for the opportunity to serve as a Guest Editor, and I appreciate the dedicated efforts of the assistant editors and reviewers in enhancing the quality of this Special Issue.

Conflicts of Interest: The author declares no conflicts of interest.

List of Contributions:

1. Zhao, Y.; Zhang, W.; Li, Y.; Fei, W.; Liang, H. Magnetic Cross-Linked Enzyme Aggregates of Glycoside Hydrolase: An Efficient and Stable Biocatalyst for Icaritin. *Catalysts* **2025**, *15*, 1034. https://doi.org/10.3390/catal15111034.
2. O'Donnell, R.; Maddaloni, M.; Scaglione, S.; Artioli, N. Post-Synthesis Ion Beam Sputtering of Pt/CeO_2–ZrO_2 Catalysts: Correlating Surface Modifications with Light-Off Performance. *Catalysts* **2025**, *15*, 1018. https://doi.org/10.3390/catal15111018.
3. Aleithan, S.H.; Laradhi, S.S.; Al-Amer, K.; El-Lateef, H.M.A. Synergistic MoS_2-Gold Nanohybrids for Sustainable Hydrogen Production. *Catalysts* **2025**, *15*, 550. https://doi.org/10.3390/catal15060550.
4. Sharaf, I.M.; Koubisy, M.S.I.; Alkallas, F.H.; Trabelsi, A.B.G.; Aboraia, A.M. Tailoring β-Bi_2O_3 Nanoparticles via Mg Doping for Superior Photocatalytic Activity and Hydrogen Evolution. *Catalysts* **2025**, *15*, 519. https://doi.org/10.3390/catal15060519.
5. Liang, J.; Li, J.; Yan, J.; Rappe, A.M.; Yang, J. Unlocking Synergistic Catalysis in NiP: Dual Role of Electronic Structure and Lewis Acidity for Enhanced Oxygen Evolution Reaction. *Catalysts* **2025**, *15*, 457. https://doi.org/10.3390/catal15050457.
6. Farghly, N.; Abu El-Oyoun, M.; Abousehly, A.; Alkallas, F.H.; Trabelsi, A.B.G.; Shaaban, E.R.; Aboraia, A.M. Investigating the Photocatalytic Properties of Reduced Graphene Oxide-Coated Zirconium Dioxide and Their Impact on Structural and Morphological Features. *Catalysts* **2025**, *15*, 289. https://doi.org/10.3390/catal15030289.
7. Kaiba, A.; Alansi, A.M.; Oubelkacem, A.; Chabri, I.; Hameed, S.T.; Afzal, N.; Rafique, M.; Qahtan, T.F. Sunlight-Driven Synthesis of $TiO_2/(MA)_2SnCl_4$ Nanocomposite Films for Enhanced Photocatalytic Degradation of Organic Pollutants. *Catalysts* **2025**, *15*, 214. https://doi.org/10.3390/catal15030214.
8. Alasmari, A.; Alresheedi, N.M.; Alzahrani, M.A.; Aldosari, F.M.; Ghasemi, M.; Ismail, A.; Aboraia, A.M. High-Performance Photocatalytic Degradation-A ZnO Nanocomposite Co-Doped with Gd: A Systematic Study. *Catalysts* **2024**, *14*, 946. https://doi.org/10.3390/catal14120946.
9. Xu, L.; Li, B.; Liu, T.; Ma, A.; Zhuang, G.; Qian, J.; Cui, Y.; Huo, S.; Xia, J.; Wang, F. Highly Efficient Degradation of 2-Methylisoborneol by Laccase Assisted by a Micro-Electric Field. *Catalysts* **2024**, *14*, 649. https://doi.org/10.3390/catal14090649.

References

1. Zlotin, S.; Egorova, K.; Ananikov, V.; Akulov, A.; Varaksin, M.; Chupakhin, O.; Charushin, V.; Bryliakov, K.; Averin, A.; Beletskaya, I. The green chemistry paradigm in modern organic synthesis. *Russ. Chem. Rev.* **2023**, *92*, RCR5104. [CrossRef]
2. Zhang, S.; Li, Z.; Wang, C.; Li, H.; Zheng, R.; He, J.; Ding, Y. Solar-Light-Driven Synergistic Photocatalytic Hydrogen Evolution and Pollutant Elimination in Aqueous Environments. *Chemsuschem* **2025**. [CrossRef] [PubMed]

3. Wang, L.; Liu, Y.; Liu, C. Green Catalysis in Nanomaterials-Photocatalysis and Electrocatalysis. *Nanomaterials* **2025**, *15*, 5. [CrossRef] [PubMed]
4. Li, R.; Feng, H.; Shi, M.; Zhang, A.; Jing, S. Expanding biocatalytic reactivity landscapes by physical fields and chemical strategies for green manufacturing. *Biotechnol. Adv.* **2026**, *86*, 108734. [CrossRef] [PubMed]
5. Bai, J.; Huang, C.; Liu, Y.; Zheng, X.; Liu, J.; Zhou, L.; Liu, J.; Jiang, Y. Integrating biocatalysis with continuous flow: Current status, challenges, and future perspectives. *J. Adv. Res.* **2025**. [CrossRef] [PubMed]
6. Zhu, Y.; Tao, L.; Chen, R.; Fu, X.-Z.; Wang, S. Mechanistic insights into atomic-to-nanoscale synergistic electrocatalysis. *Sci. China-Mater.* **2025**, *68*, 2587–2600. [CrossRef]
7. Harada, S.; Tsuji, N.; Fukushima, S.; Yamamoto, J.; Arai, S.; Nemoto, T. Redox-Adaptive Auto-Tandem Catalysis: Ce(III)/Ce(IV) Interconversion-Mediated Integration of Nazarov Cyclization and Oxidative Hydroxylation. *Acs Catal.* **2025**, *15*, 14341–14347. [CrossRef]
8. Liu, K.; Liao, Y.; Xiao, Q.; He, W.; He, F.; Jiang, H.; Chen, Q.; Chang, Z.; Hou, S.; Wang, H. All-day extraction of uranium by photo-triggered uranium self-catalysis coupled with $(UO_2)O_2{\bullet}4H_2O/Bi_4NbO_8Cl$ piezo-catalysis. *J. Environ. Chem. Eng.* **2025**, *13*, 118508. [CrossRef]
9. Shi, K.; Meng, H.; Liu, J.; Ma, S.; Tang, W. Synergistic Contact-Electro-Catalysis and Photocatalysis via TiO_2@PTFE Composites for Efficient N_2 to NH_3 Conversion. *Angew. Chem.-Int. Ed.* **2025**, *64*, e202515707. [CrossRef] [PubMed]
10. Yang, X.; Huang, J.; Guo, J.; Fang, S.; Wang, Z.; Wu, G.; Wu, Y.; Zhong, F. Bridging chemistry and biology for light-driven new-to-nature enantioselective photoenzymatic catalysis. *Chem. Soc. Rev.* **2025**, *54*, 5157–5188. [CrossRef] [PubMed]
11. Pan, D.; Wang, Y.; Liang, Q.; Zhou, M.; Li, X.; Xu, S.; Li, Z. Recent advances in solar-driven photothermal nanostructured materials for CO_2 reduction: A review. *J. Environ. Chem. Eng.* **2023**, *11*, 110324. [CrossRef]

catalysts

MDPI

Article

Highly Efficient Degradation of 2-Methylisoborneol by Laccase Assisted by a Micro-Electric Field

Ling Xu [1,†], Beidian Li [1,†], Tingting Liu [1], Anzhou Ma [2], Guoqiang Zhuang [2,3], Jingya Qian [1], Yi Cui [1], Shuhao Huo [1], Jiexiang Xia [4] and Feng Wang [1,*]

1 School of Food and Biological Engineering, Jiangsu University, Zhenjiang 212013, China; lxu@ujs.edu.cn (L.X.); 2222318044@stmail.ujs.edu.cn (B.L.); 2221918053@stmail.ujs.edu.cn (T.L.); sjqjy@126.com (J.Q.); cuiyi@ujs.edu.cn (Y.C.); huo@ujs.edu.cn (S.H.)

2 Key Laboratory of Environmental Biotechnology, Research Center for Eco-Environmental Sciences, Chinese Academy of Sciences, Beijing 100085, China; azma@rcees.ac.cn (A.M.); gqzhuang@rcees.ac.cn (G.Z.)

3 School of Energy and Environment, Binzhou Institute of Technology, Binzhou 256606, China

4 School of Chemistry and Chemical Engineering, Jiangsu University, Zhenjiang 212013, China; xjx@ujs.edu.cn

* Correspondence: fengwang@ujs.edu.cn

† These authors contributed equally to this work.

Abstract: Taste and odor (T&O) compounds have emerged as crucial parameters for assessing water quality. Therefore, identifying effective methodologies for the removal of these compounds is imperative. In this study, an effective approach utilizing laccase assisted by a micro-electric field was developed for the degradation of 2-methylisoborneol (2-MIB). For this purpose, the optimal conditions for the laccase-catalyzed degradation of 2-MIB were determined, and they were pH 4.0, 25 °C, 150 rpm, 0.1 U/mL of laccase, and 200 ng/L of 2-MIB. Under these specified conditions, the degradation efficiency of 2-MIB was approximately 78% after a 4 h reaction period. Subsequently, the introduction of an electric field yielded a synergistic effect with the enzyme for 2-MIB degradation. At an electric current intensity of 0.04 A over a 4 h duration, the degradation efficiency increased to 90.78%. An analysis using SPME-GC/MS provided information on the degradation intermediates of 2-MIB resulting from laccase-catalyzed degradation, electrocatalytic degradation, and micro-electric-assisted laccase degradation. The potential degradation pathways of 2-MIB illustrated that these three methods result in common degradation products, such as capric aldehyde, nonylaldehyde, and 2-ethylhexanol, and their final products include 3-pentanone, acetone, and 2-butanone. This study provides an enzyme–electrochemical method for the efficient and rapid degradation and removal of 2-MIB. The strategy of laccase catalysis assisted by a micro-electric field has good potential for the removal of pollutants from the natural environment.

Keywords: 2-methylisoborneol; laccase; micro-electric field; oxidation degradation; possible degradation pathway

1. Introduction

Taste and odor (T&O) compounds not only jeopardize the safety of drinking water but also increase the operational costs of water treatment processes. These compounds can significantly impact the palatability of water, prompting consumer dissatisfaction and necessitating additional measures to ensure water quality and safety [1]. T&O compounds originate from two primary sources. One source is the improper disposal of water resources by individuals, which can introduce odoriferous substances into natural water bodies, including chemicals, mineral salts, inorganic nitrides, and sulfides. The other source is excretions from algae, actinomycetes, fungi, and organic materials such as humus, which contribute diverse odors to water [2,3]. Various T&O compounds are present in natural water, such as geosmin (GSM), 2-methylisoborneol (2-MIB), 2-isopropyl-3-methoxypyrazine (IPMP), 2-isobutyl-3-methoxypyrazine (IBMP), and 2,4,6-trichloroanisole (TCA) [2–4]. Among these compounds, 2-MIB is particularly noteworthy as a significant odorant in drinking water sources, characterized by a musty, earthy odor primarily originating from blue-green algae and actinomycetes [5].

2-MIB is considered a typical T&O compound due to its exceedingly low odor detection threshold in the ng/L range [6], which poses significant challenges to its effective removal using conventional water treatment methods [3,7]. While T&O substances in drinking water typically do not pose a direct health threat to humans, their presence can lead individuals to reject water due to the undesirable taste, potentially resulting in refusal to consume or utilize it. Consequently, the removal of 2-MIB from drinking water has garnered significant attention in recent years. The removal of 2-MIB from water involves a combination of physical, chemical, and biological methods [3,8,9]. Traditional physical water treatment techniques such as coagulation, precipitation, sand filtration, and activated carbon adsorption have exhibited minimal efficacy and notable instability in their capacity to eliminate these compounds [10]. Likewise, conventional chemical oxidation processes such as chlorination and permanganate oxidation have not only demonstrated low efficiency but also possess a limited capability to oxidize odorous compounds, potentially yielding hazardous by-products [11,12]. Biological methods have shown greater promise for the removal of 2-MIB, with studies indicating that specific aquatic and soil bacteria can effectively biodegrade this compound [13,14]. Bacteria are crucial in the degradation of 2-MIB, breaking down, co-metabolizing, and enzymatically degrading the compound through diverse metabolic processes [15]. Additionally, enzymes isolated from 2-MIB degraders have been shown to effectively reduce 2-MIB concentrations [16].Nevertheless, research into enzymatic degradation methods for 2-MIB remains limited, underscoring the significant potential of enzymatic approaches in tackling the challenge of 2-MIB removal.

Laccase, a widely distributed multicopper glycoprotein enzyme, is renowned for its capability to oxidize a diverse range of both phenolic and non-phenolic compounds. Enzymatic degradation utilizing laccase has garnered significant attention due to its high removal efficiency and environmentally friendly characteristics. This enzyme has demonstrated the ability to oxidize a broad spectrum of substrates, encompassing environmental contaminants such as pharmaceuticals, dyes, pesticides, phenols, and components commonly found in wastewater [17–19]. Enhancing laccase catalytic efficiency through the use of mediators and immobilization techniques and exploring laccase from diverse sources are promising strategies to improve the bioremediation of environmental pollutants [20]. These strategies can enhance the performance of laccase in degrading various pollutants, positioning it as a promising tool for environmental remediation efforts.

Electrochemical water treatment technology utilizes an applied potential or current to facilitate electron transfer and multi-proton-coupled electron transfer processes [21]. This process leads to the oxidation, reduction, adsorption, and migration of pollutants [22]. Electrocatalysis has sparked significant interest in the field of electrochemical oxidation [23]. On the one hand, it follows non-radical pathways, involving direct electron transfer processes, where pollutants act as electron donors or acceptors, leading to direct oxidation or reduction [24]. On the other hand, it can generate powerful oxidants, such as hydroxyl radicals, at the catalytic electrode surface through radical pathways. These oxidants effectively oxidize and decompose pollutants, leading to efficient water purification. Most reviews on electrocatalysis technology have focused on specific areas, with some studies exploring the co-catalytic oxidation of various oxidants and peroxides [25].

This study aimed to explore how different reaction conditions affect the free laccase-catalyzed degradation of 2-MIB, optimize these conditions, and investigate the mechanism under optimal settings. To enhance electron transfer during the laccase-catalyzed oxidation process, this research examined the influence of an external micro-electric field on 2-MIB degradation by laccase. This study compared the efficiency of 2-MIB degradation catalyzed by laccase alone, a micro-electric field alone, and micro-electric-field-assisted laccase. Additionally, the potential degradation pathways of 2-MIB were also analyzed.

2. Results and Discussion

2.1. Effects of Reaction Parameters on Degradation of 2-MIB Catalyzed by Laccase

2.1.1. Effect of pH on Catalytic Degradation of 2-MIB by Laccase

At pH 4.0, the degradation efficiency of 2-MIB was optimal, achieving a degradation efficiency of 77.80% (Figure S1). Deviations from this optimal pH significantly reduced the degradation efficiency of 2-MIB, particularly under alkaline conditions. For instance, at pH 8.0, laccase catalyzed the degradation of 2-MIB at only 27.46% of the rate observed at pH 4.0. The efficacy of the biocatalysts was notably compromised under both acidic and alkaline conditions, with a more pronounced effect observed in alkaline environments. This was because pH influenced not only the substrate's ionization but also the spatial arrangement of the enzyme molecules and the ionization state of their surface-active groups [26]. Under alkaline conditions, the diminished degradation efficiency of 2-MIB could be attributed to the strong electrostatic repulsion between the oppositely charged groups at the enzyme's active site, leading to structural alterations that impacted its functionality [27].

According to the results in Figure 1a, linear fitting of the degradation efficiency plot was performed (Figure 1a), and the apparent rate constant for 2-MIB was calculated (Figure 1b). As shown in Figure 1b, the rate constant (k) at pH 4.0 was 0.0091 min^{-1}, representing the maximum reaction rate. Deviations from pH 4.0, either lower or higher, resulted in a decreased reaction rate. Laccase exhibited higher biological activity at the optimal pH, as different pH levels affected the charged states of both the substrate and the enzyme, influencing their binding interaction [28]. In this reaction, laccase primarily served as a catalyst, and changes in pH impacted its activity, thereby affecting the overall reaction rate. Similarly, Onaizi and Alshabib also found a non-linear trend in the conversion rate of bisphenol A catalyzed by laccase, where the rate first increased and then decreased with rising pH levels, indicating an optimal pH value [29].

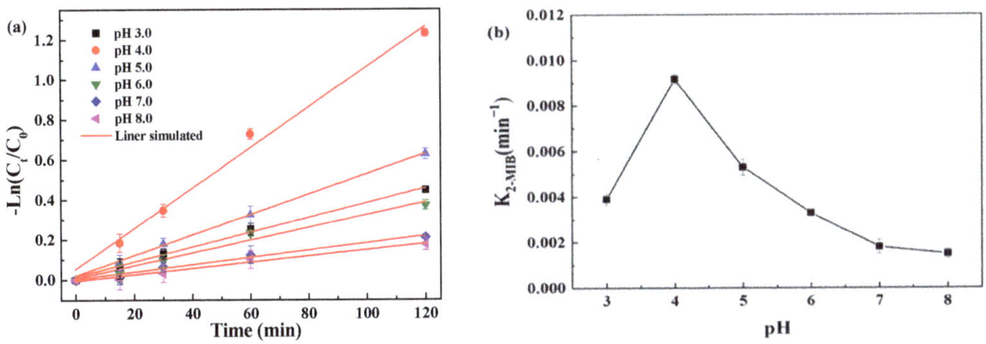

Figure 1. Fitting linear plots (**a**) and apparent rate constants (**b**) of 2-MIB degradation efficiency at different pH values (laccase dosage: 0.1 U/mL, 2-MIB concentration: 200 ng/L, 25 °C, 150 r/min).

2.1.2. Effect of Temperature on Catalytic Degradation of 2-MIB by Laccase

As shown in Figure S2, the degradation efficiency of 2-MIB reached 78.24% at 25 °C, indicating optimal efficiency. When the temperature exceeded 25 °C, the degradation efficiency of 2-MIB significantly decreased, dropping to 37.84% at 40 °C. Temperature affected the degradation of 2-MIB in two main ways: Firstly, the degradation efficiency of 2-MIB catalyzed by laccase increased with respect to the temperature. This was because the enzymes had an optimal temperature for their best activity, and the enzyme activity was suppressed at lower temperatures [30,31]. Secondly, the degradation efficiency of 2-MIB decreased when the temperature increased to more than 25 °C; this could be due to the heat-induced conformational changes in the enzyme at higher temperatures, leading to enzyme inactivation [29]. Based on the results in Figure S2, linear fitting of the degradation efficiency plot was performed (Figure 2a), and the apparent rate constant of 2-MIB was calculated (Figure 2b). At 25 °C, the k value was 0.0089 min^{-1}, reaching the maximum reaction rate. In the ascending region below the optimal temperature, the increase in the reaction rate constant due to the rising temperature was moderate. Conversely, in the descending region above the optimal temperature, the rate of enzyme deactivation due to heat became more significant, leading to a substantial decrease in the efficiency of 2-MIB degradation. It was reported that the reaction rate for the degradation of polycyclic aromatic hydrocarbons catalyzed by laccase exhibited an initial incline followed by a decline as the temperature increased, consistent with the results of this study [32].

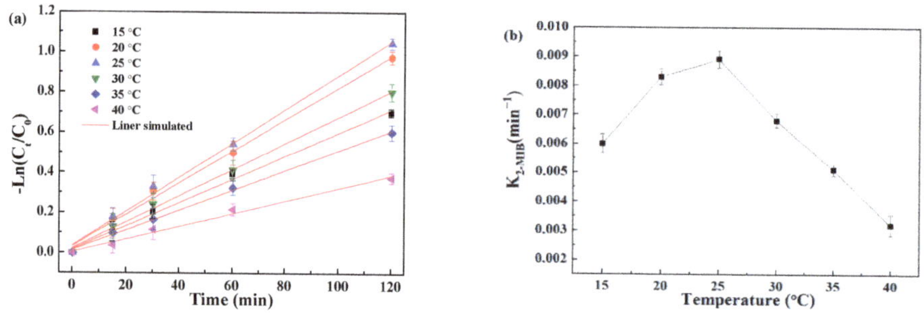

Figure 2. Fitting linear plots (**a**) and apparent rate constants (**b**) of 2-MIB degradation efficiency at different temperatures (laccase dosage: 0.1 U/mL, 2-MIB concentration: 200 ng/L, pH 4.0, 150 r/min).

2.1.3. Effect of Stirring Speed on Catalytic Degradation of 2-MIB by Laccase

Upon increasing the agitation speed from 75 r/min to 150 r/min, the degradation efficiency of 2-MIB increased from 40.72% to 78.28% (Figure 3a). Agitation served to improve the mixing of the substrate and enzyme in the reaction system, enhancing their interaction frequency and thereby facilitating the reaction [33]. At lower rotational speeds, the rate of the entire process is governed by mass transfer, whereas increasing the agitation speed consistently enhances the mass transfer rates, thereby boosting degradation efficiency [34]. However, further escalation of the agitation speed to 200 r/min resulted in a decline in the degradation efficiency of 2-MIB. In the degradation of ciprofloxacin hydrochloride catalyzed by laccase assisted by ultrasound, a higher stirring rate resulted in worse degradation, which is similar to the trend observed in this study [35]. Therefore, an agitation speed of 150 r/min was chosen for the optimal degradation of 2-MIB. Based on the results in Figure S3, linear fitting of the degradation efficiency plot was performed (Figure 3a), and the apparent rate constant of 2-MIB was calculated (Figure 3b). In Figure 3b, it is evident that, with an increasing agitation speed, the reaction rate initially increased and then decreased, peaking at 150 r/min with a k value of 0.0081 min^{-1}.

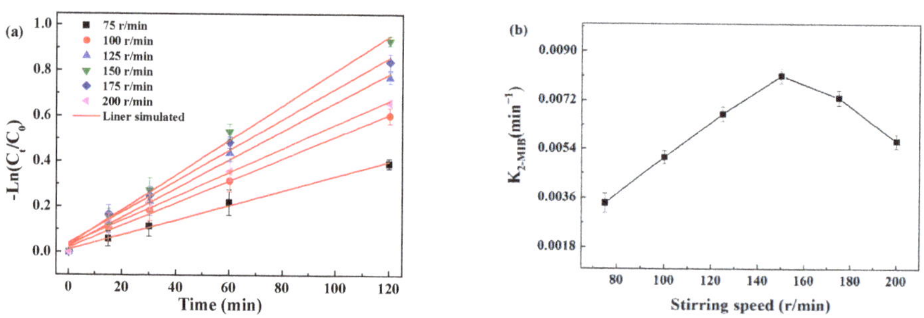

Figure 3. Fitting linear plots (**a**) and apparent rate constants (**b**) of 2-MIB degradation efficiency at different stirring speeds (laccase dosage: 0.1 U/mL, 2-MIB concentration: 200 ng/L, 25 °C, pH 4.0).

2.1.4. Effect of Enzyme Dosage on Catalytic Degradation of 2-MIB by Laccase

As shown in Figure S4, when the concentration of laccase was 0.01 U/mL, the degradation efficiency of 2-MIB after 4 h was only 25.69%. An insufficient enzyme dosage likely reduced the frequency of substrate–enzyme interactions, thereby hindering the degradation of 2-MIB [36]. Increasing the laccase concentration to 0.1 U/mL significantly boosted the degradation efficiency to the maximum of 78.24%. The optimal enzyme concentration enhanced the reaction rate by maximizing the frequency of substrate–enzyme interactions, thereby facilitating the degradation of 2-MIB. In the case of 0.1 U/mL, the concentration of laccase was 1.3 nM, and the concentration of 2-MIB was 1.2 nM. However, further increases in the laccase concentration to 0.15 U/mL and 0.2 U/mL did not significantly enhance the degradation efficiency beyond that achieved at 0.1 U/mL. At a higher enzyme concentration, the binding of the substrate with the enzymes was saturated, and the enzyme was redundant [37,38]. Therefore, a laccase concentration of 0.1 U/mL appeared to be optimal for efficiently degrading 200 ng/L of 2-MIB.

Based on the results in Figure S4, linear fitting of the degradation efficiency plot was performed (Figure 4a), and the apparent rate constant of 2-MIB was calculated (Figure 4b). It is evident in Figure 4a that increasing the enzyme dosage from 0.01 U/mL to 0.1 U/mL resulted in a significant increase in the k value to 0.0081 min^{-1}, thereby substantially enhancing the reaction rate of 2-MIB. However, further increasing the enzyme dosage to 0.2 U/mL only slightly raised the k value from 0.0083 min^{-1} to 0.0097 min^{-1}, representing a minor increase of 0.0014 min^{-1}, with no significant improvement in the reaction rate of 2-MIB. Therefore, the enzymatic dosage of 0.1 U/mL was selected as the appropriate condition for 2-MIB degradation.

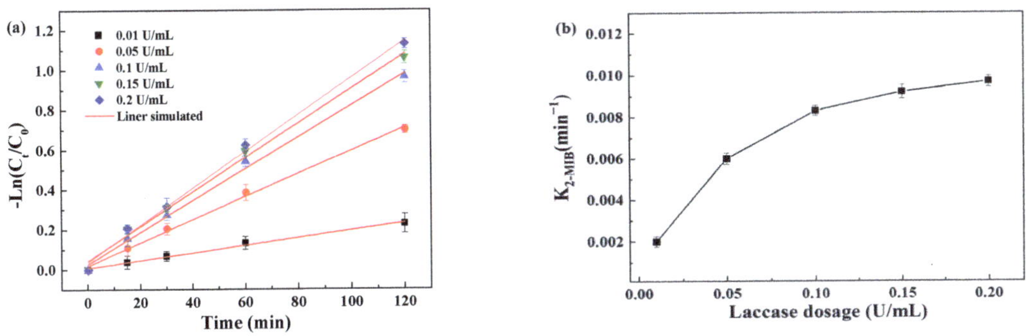

Figure 4. Fitting linear plots (**a**) and apparent rate constants (**b**) of 2-MIB degradation efficiency at different laccase dosages (2-MIB concentration: 200 ng/L, 25 °C, pH 4.0, 150 r/min).

2.1.5. Effect of Initial 2-MIB Concentration on Its Degradation Catalyzed by Laccase

In Figure S5, as the initial concentration of 2-MIB increased from 50 ng/L to 200 ng/L, the degradation efficiency rose from 51.49% to 78.34%. This was due to the enhanced opportunities for laccase–2-MIB interactions at higher concentrations, thereby augmenting degradation. However, when the initial concentration was further elevated to 500 ng/L, the degradation efficiency declined to 69.14%. Although the degradation efficiency was the highest at 200 ng/L, the actual degradation amount of 2-MIB increased with respect to the substrate concentration. These results indicate that, within a certain range, increasing the 2-MIB concentration could enhance the reaction rate, but high concentrations may negatively impact degradation efficiency [39,40]. This decrease was attributed to the limited availability of active sites on laccase at higher 2-MIB concentrations. Additionally, the higher initial 2-MIB concentration resulted in an increased production of degradation by-products, which may compete with the substrate for the binding sites of laccase or inhibit enzyme activity, thus reducing the overall degradation efficiency [41]. In the case of 500 ng/L, the concentration of laccase was 1.3 nM, and the concentration of 2-MIB was

3.0 nM. Based on the results in Figure S5, linear fitting of the degradation efficiency was performed (Figure 5a), and the apparent rate constant of 2-MIB was calculated (Figure 5b). It is evident in Figure 5b that, with an increase in the initial 2-MIB concentration, the reaction rate initially increased and then decreased. The peak reaction rate was observed at an initial 2-MIB concentration of 200 ng/L, with a rate constant (k) of 0.0089 min^{-1}. At the same time, under our reaction conditions, we found that there was no loss of enzyme activity after 4 h. In order to better understand the data, the kinetic parameters of laccase in the catalysis of 2-MIB were determined (Figure S6), where K_m was 3.75 mmol/L, and K_{cat} was 1.27×10^3 (Table S1).

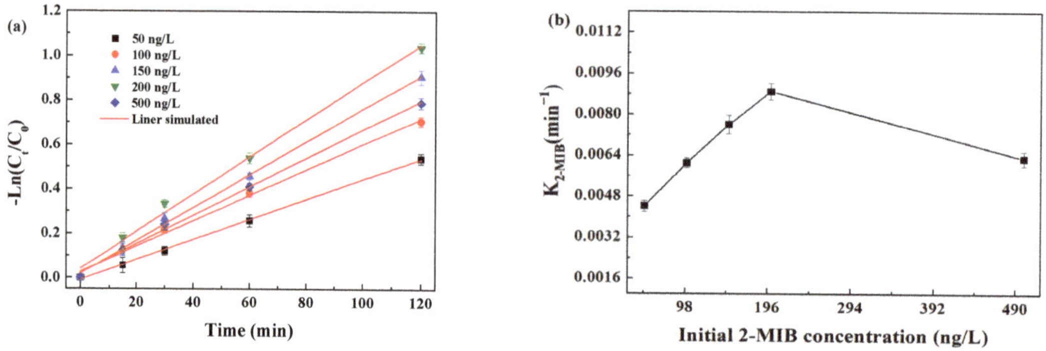

Figure 5. Fitting linear plots (**a**) and apparent rate constants (**b**) of 2-MIB degradation efficiency at different initial 2-MIB concentrations (laccase dosage: 0.1 U/mL, 25 °C, pH 4.0, 150 r/min).

2.1.6. Effect of Electric Field Intensity on 2-MIB Degradation Catalyzed by Laccase

The catalytic mechanism of laccase involved sequential stages: substrate binding, electron transfer resulting in the formation of intermediates and reduction of the intermediates [40]. Research has highlighted that the rate-limiting step in the oxidation of various substrates by laccases is the transfer of substrate electrons to the type I copper (Cu) center of the enzyme's native intermediate [42,43]. The impact of a direct current electric field on laccase activity in aqueous systems was investigated, and an enhancement in laccase activity was observed near the positive electrode, resulting in the acceleration of the catalytic reaction rate [44]. This study's results showed that the application of an external electric field increased the degradation efficiency of 2-MIB catalyzed by laccase. As the applied current increased from 0.01 A to 0.04 A, the degradation efficiency of 2-MIB increased to 90.78% (Figure 6). This was likely due to the enhanced electron transfer between the electrode, solution, and laccase with the increased current. The higher current generated a large number of highly oxidative free radicals, thereby promoting the degradation of 2-MIB [45]. However, an incremental gain in degradation efficiency occurred when the current increased from 0.04 A to 0.08 A. In fact, the application of an external electric field induced a gradual rise in pH in the cathode region (pH 4.2) and a decline in pH in the anode region (pH 3.8). Thus, the anode region became marginally acidic, while the cathode region became more alkaline with a higher current, influencing laccase activity. As depicted in Figure 7, the application of a micro-electric field alone in the degradation of 2-MIB provided a degradation efficiency of only 8.15% after a 4 h reaction. In contrast, the degradation catalyzed by laccase resulted in a degradation efficiency of 77.65% for 2-MIB. These results indicate that the degradation efficiency of 2-MIB catalyzed by laccase assisted by a micro-electric field was higher than the sum of the degradation efficiencies obtained using a micro-electric field and laccase. Therefore, the catalytic reaction by laccase assisted by a micro-electric field exhibited a synergistic effect, which could be due to the electric field's acceleration of electron transfer [44,46], facilitating the laccase-catalyzed reaction. The combination of an electric field and bioremediation was used to degrade aliphatic hydrocarbons and achieved good degradation efficiency [47]. The application of an appropriate direct current can stimulate microbial activity and accelerate the biodegradation of petroleum pollutants [48]. A large-scale electro-biological method was utilized to repair soil contaminated by PAHs, where the degradation degree of PAHs was significantly enhanced compared with that achieved by bioremediation alone [49]. These results indicate that the combined strategy of an electric field and biological treatment has good potential for the removal of pollutants.

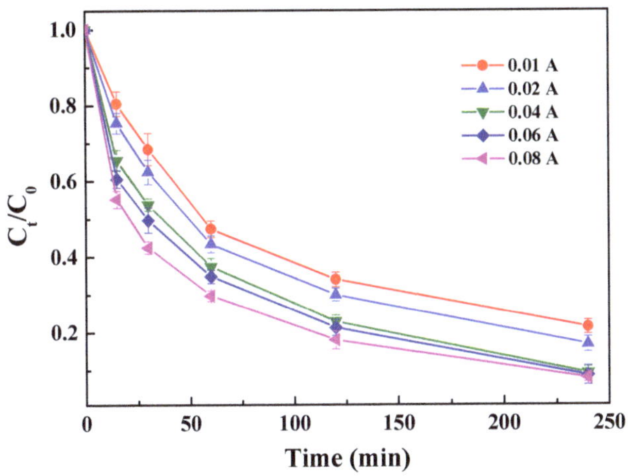

Figure 6. Time profile of 2-MIB degradation by laccase at different electric field intensities.

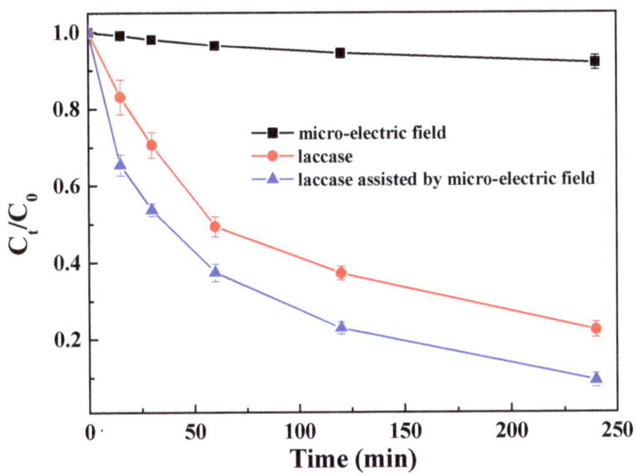

Figure 7. Time profile of 2-MIB degradation by micro-electric field, laccase, and laccase assisted by a micro-electric field (current intensity: 0.04 A, laccase dosage: 0.1 U/mL, 2-MIB concentration: 200 ng/L).

2.2. *Analysis for the Intermediates of 2-MIB Degradation*

Through an SPME-GC/MS analysis, the degradation intermediates of 2-MIB resulting from different reaction processes were identified and are listed in the Supplementary Materials (Tables S1–S3). Based on the information, the potential degradation pathways of 2-MIB catalyzed using different strategies were deduced and are illustrated in Figures 8–10.

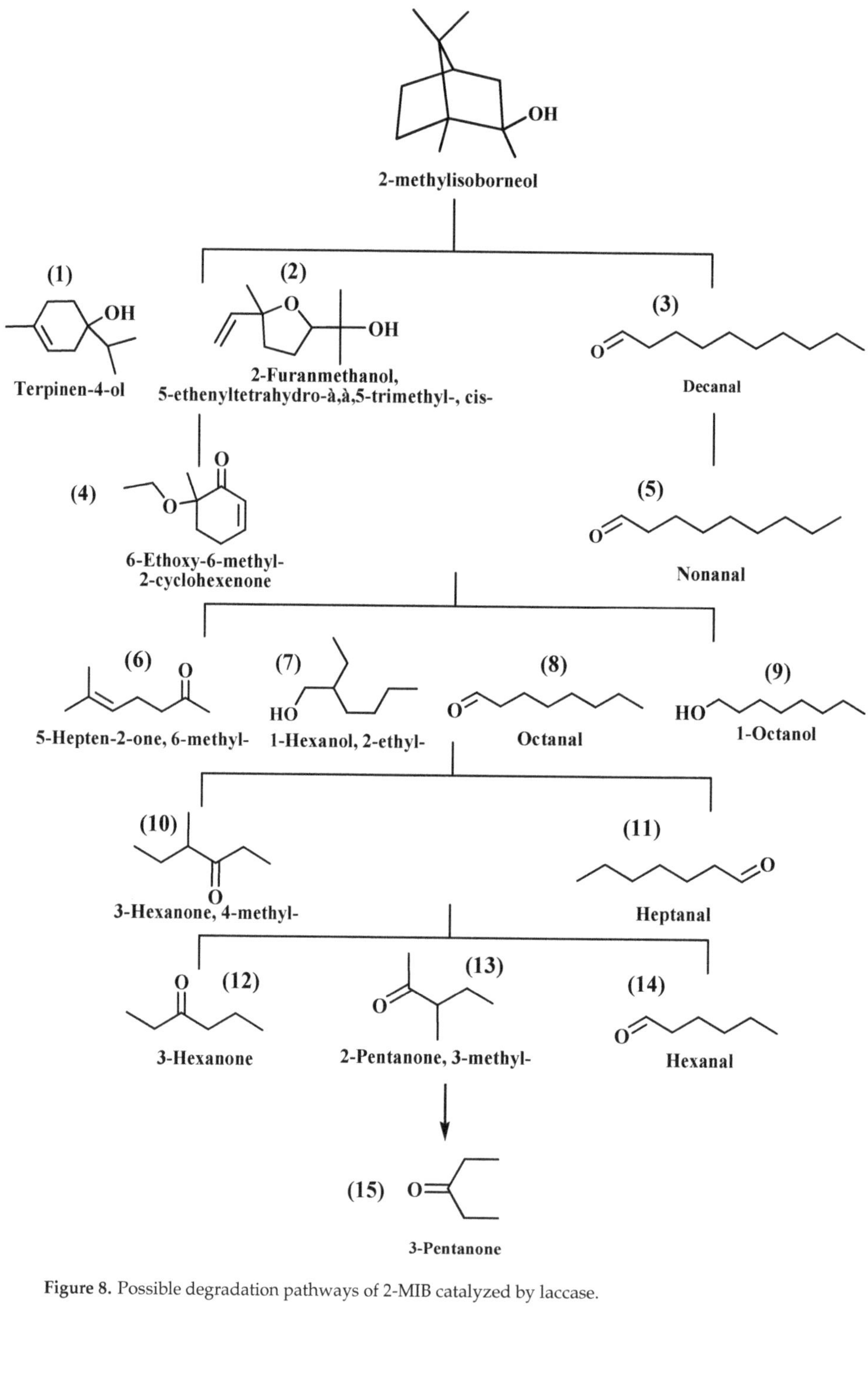

Figure 8. Possible degradation pathways of 2-MIB catalyzed by laccase.

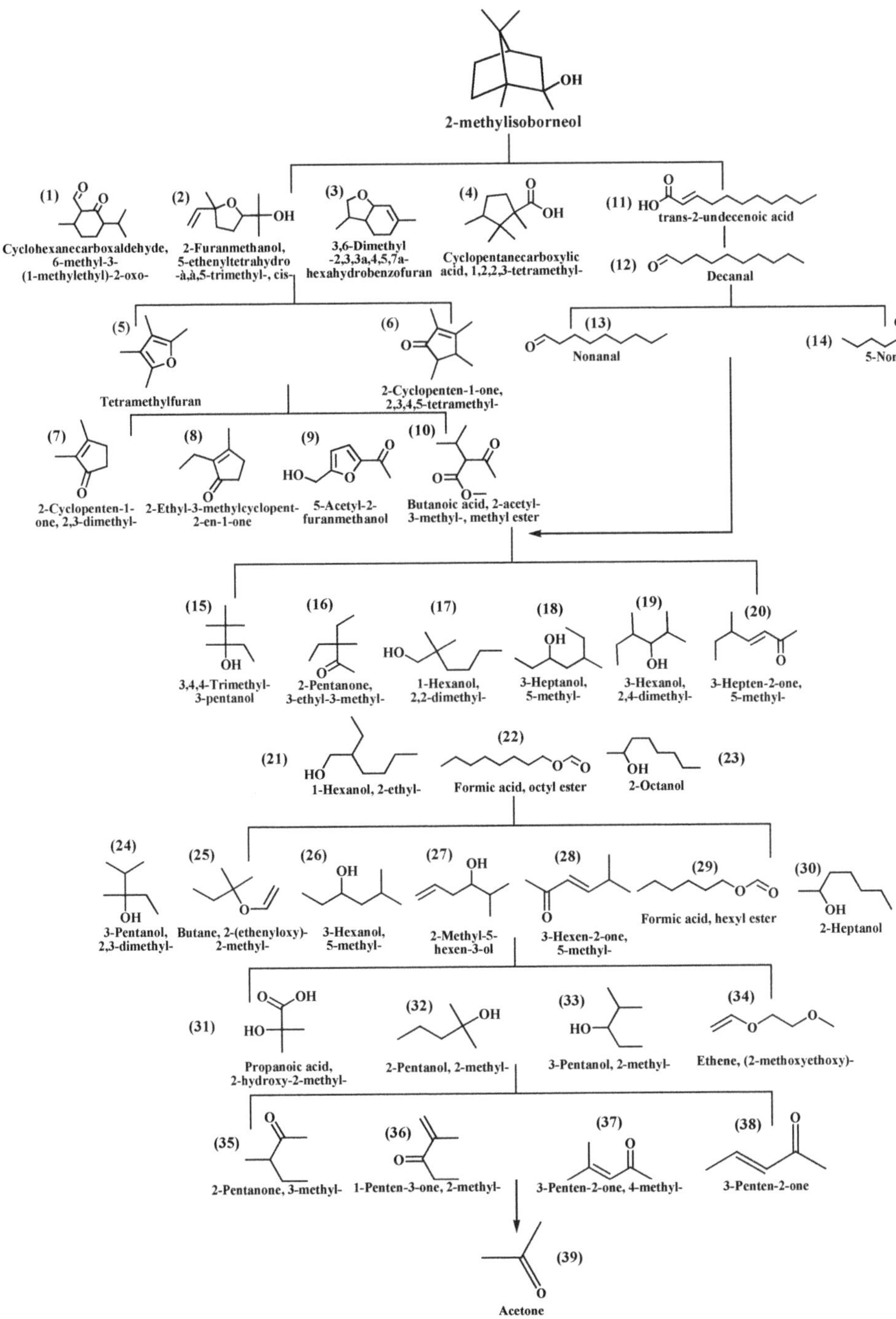

Figure 9. Possible degradation pathways of 2-MIB catalyzed by a micro-electric field.

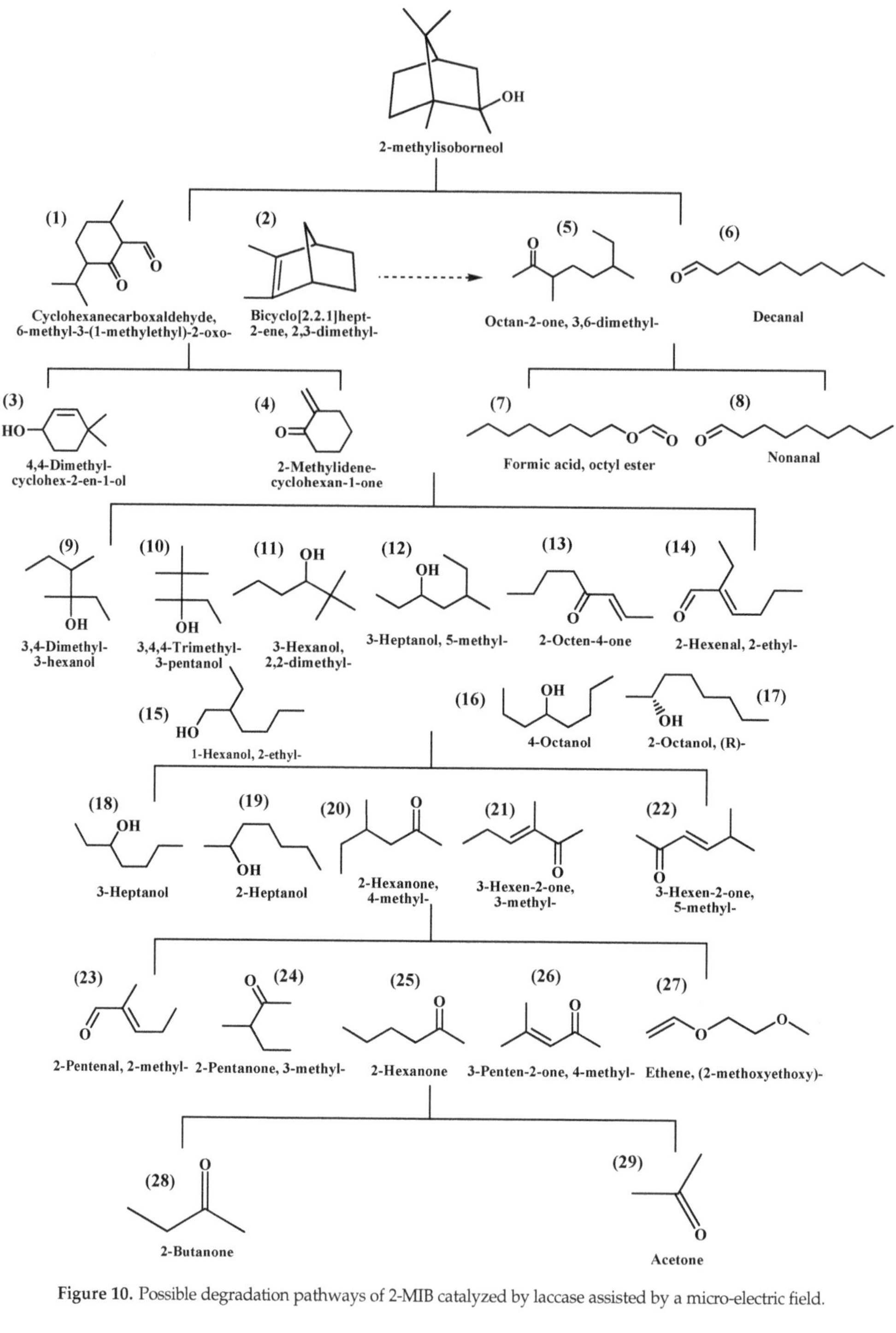

Figure 10. Possible degradation pathways of 2-MIB catalyzed by laccase assisted by a micro-electric field.

2.2.1. Possible Degradation Pathway of 2-MIB Catalyzed by Laccase

Figure 8 depicts the potential degradation pathways of 2-MIB catalyzed by laccase under previously optimized conditions. Products (**1**) and (**2**), comprising ten carbons, were likely formed through the oxidation of 2-MIB. Further oxidation by laccase may have led to the formation of the nine-carbon monocyclic compound 6-ethoxy-6-methyl-2-cyclohexenone (Product **4**). Alternatively, 2-MIB may have undergone ring-opening and recombination reactions, directly yielding the ten-carbon linear compound decanal (Product **3**). Due to the instability of linear chains relative to branched structures, further C-C bond cleavage resulted in the formation of the nine-carbon linear compound nonanal (Product **5**). Products (**4**) and (**5**), each containing nine carbons, underwent additional oxidation or monocyclic cleavage to yield four linear compounds with eight carbons: 6-methyl-5-hepten-2-one, 2-ethyl-1-hexanol, octanal, and 1-octanol (Products **6–9**). This degradation pathway involved oxidation, ring-opening, and recombination reactions, ultimately yielding a shorter-chain compound with fewer carbons (Product **15**), namely, 3-pentanone. Enzymes from *Pseudomonas mandelii* were used for 2-MIB degradation, and the final degradation product of 2-methyl-2-bornene was detected [16]. In the catalytic degradation process of cyclic compounds (bisphenol A) using laccase as a catalyst, similar reactions occurred, including oxidation and ring-opening reactions [50].

2.2.2. Possible Degradation Pathways of 2-MIB by Electrocatalysis

Figure 9 shows the possible degradation pathways of 2-MIB under a current of 0.04 A. Initially, 2-MIB underwent oxidation processes generating a spectrum of five-membered rings, six-membered rings, and bicyclic compounds (**1–6**). These intermediates subsequently underwent further oxidation or demethylation to yield three five-membered rings (**7–9**) and ring-opening compounds (**10**). Further oxidation or C-C bond cleavage resulted in the formation of diverse alcohols, ketones, esters, and other intricate branched-chain compounds. This degradation pathway encompassed reactions such as oxidation, ring-opening, and cleavage reactions, ultimately resulting in the production of the small molecule acetone. In contrast, previous investigations into the electrocatalytic degradation of 2-MIB have observed its ultimate transformation into carbon dioxide and water [50]. Notably, this distinction may be attributed to the elevated current density employed in those studies (2.5 mA cm^{-2}), markedly differing from the lower current density of 0.08 mA cm^{-2} utilized in this study, thereby yielding different final degradation products [51].

The chemical transformations that occurred in the presence of an electric field were notably complex. Typically, the degradation of pollutants under electrocatalytic conditions primarily proceeded through anodic oxidation, which encompassed both direct oxidation reactions and indirect oxidation mediated by hydroxyl radicals [52]. Direct oxidation entailed the potential adsorption of pollutants onto the electrode surface, where they interacted with active oxygen species adsorbed on the electrode [52,53]. Indirect oxidation, facilitated by hydroxyl radicals, involves generating these radicals through the oxidation of water on the anode surface under the influence of the electric field. These hydroxyl radicals subsequently participated in the oxidation of pollutants [52]. Guided by the principles of electrocatalysis, the degradation of 2-MIB involved the cleavage of chemical bonds and electron transfer processes, leading to the gradual transformation of the compound into lower-molecular-weight intermediates.

2.2.3. Possible Degradation Pathways of 2-MIB Catalyzed by Laccase Assisted by a Micro-Electric Field

Figure 10 integrates the potential degradation pathways of 2-MIB via laccase-catalyzed degradation assisted by a micro-electric field. Initially, 2-MIB underwent oxidation to form cyclic compounds (**1**) or dehydrogenation to yield compounds (**2**). These intermediates could be further oxidized to two cyclic compounds, namely, 4,4-Dimethyl-cyclohex-2-en-1-ol (**3**) and 2-Methylidene-cyclohexan-1-one (**4**), both comprising two cyclic units. Alternatively, ring-opening reactions may have generated branched ketone compounds containing

10 carbons, such as Octan-2-one, 3,6-dimethyl- (**5**). Subsequently, C-C bond cleavage and successive oxidation led to the formation of shorter-chain compounds. Moreover, 2-MIB underwent ring-opening and recombination reactions, directly forming long-chain compounds with 10 carbons (**5**, **6**). These long-chain compounds, unstable in structure, further underwent C-C bond cleavage to generate ketones, alcohols, and esters with nine or eight carbons. This degradation process involved oxidation, dehydrogenation, and ring-opening reactions, ultimately resulting in the formation of the small-molecule compounds 2-butanone (**28**) and acetone (**29**).

By comparing these three degradation pathways, it was observed that 2-MIB degradation generated some common degradation products, such as decanal, nonanal, and 2-ethylhexanol. Additionally, while electro-assisted enzyme catalysis accelerated the degradation of 2-MIB by promoting enzyme-catalyzed reactions, the enhancement effects were more complex, resulting in a greater variety of degradation intermediates than enzymatic degradation alone.

3. Materials and Methods

3.1. Chemicals

The 2-MIB standard (98% purity) was procured from Wako Pure Chemicals, Ltd. (Osaka, Japan). It was dissolved in methanol to prepare as a stock solution with a concentration of 20 mg/L. Laccase derived from *Trametes versicolor*, with an enzymatic activity of ≥0.5 U/mg, was obtained from Sigma-Aldrich Corporation (St. Louis, MO, USA). All other materials were of analytical grade and were provided by Sinopharm Chemical Reagent Co., Ltd. (Shanghai, China).

3.2. Experimental Procedure for 2-MIB Degradation

3.2.1. Preparation of Laccase Stock Solution

A specific amount of laccase was dissolved in 0.1 mol/L pH 3.5 acetic acid–sodium acetate buffer solution. After stirring for 30 min, the solution was filtrated through a 0.22 μm filter and stored at 4 °C until further experimentation.

3.2.2. Degradation of 2-MIB Catalyzed by Laccase

The degradation experiment was carried out in a 2-MIB (200 ng/L) solution with 0.1 U/mL of laccase. The experiment was conducted in a shaker at 25 °C and 150 r/min. Samples were taken during the reaction, and a certain amount of 6 mol/L NaOH solution was added to the samples to stop the reaction. Finally, headspace solid-phase microextraction gas chromatography–mass spectrometry (SPME-GC/MS) was used to determine the residual 2-MIB in the samples.

The effects of pH on the degradation efficiency and kinetics of 2-MIB were studied by adjusting the initial pH to different levels (3.0, 4.0, 5.0, 6.0, 7.0, 8.0). To determine the influence of temperature and shaking speed on 2-MIB degradation, different temperatures (15, 20, 25, 30, 35, 40 °C) and various shaking speeds (75, 100, 125, 150, 175, 200 r/min) were tested. The effects of laccase dosage (0.01, 0.05, 0.1, 0.15, 0.2 U/mL) and initial 2-MIB concentration (50, 100, 150, 200, 500 ng/L) on degradation were also investigated.

3.2.3. Degradation of 2-MIB Catalyzed by Micro-Electric Field

The degradation of 2-MIB catalyzed by a micro-electric field with/without laccase was carried out in the electrocatalysis device described in our previous research with some modifications [54]. For this test, graphite–graphite plate electrodes were used, and a magnetic agitator was applied to ensure mixing during catalysis. For the degradation of 2-MIB, the optimized conditions obtained in Section 3.2.2 were used.

3.3. Analysis

3.3.1. Determination of Laccase Activity

An appropriate amount of laccase was added to 3 mL of 0.1% (*w*/*v*) catechol solution prepared in 0.1 mol/L pH 3.0 tartaric acid buffer. The reaction was carried out at 25 °C for 5 min, and the change in absorbance at 450 nm was measured. The amount of laccase required to oxidize one micromole of catechol per minute was defined as one enzyme activity unit (U).

3.3.2. Analysis of 2-MIB and Its Intermediate Degradation Products

The analysis and identification of 2-MIB and its potential degradation products were carried out using SPME-GC/MS. The headspace extraction time was 50 min, and the injection was performed manually. The chromatographic column used was HP-5MS (30 m × 0.25 mm × 0.25 µm). Helium was used as the carrier gas at a flow rate of 1 mL/min. The injection temperature was set to 270 °C. The column temperature was initially maintained at 60 °C for 1 min, then increased to 150 °C at a rate of 5 °C/min, and finally increased to 270 °C at a rate of 20 °C/min and maintained for 5 min. The mass spectrometry conditions were set as follows: the ion source temperature was 280 °C, the quadrupole temperature was 300 °C, the ionization mode was Electron Impact (EI), and the mass range was 50 *m*/*z* to 300 *m*/*z*. The mass spectra of all detected compounds were compared with those in NIST spectral library 2.0.

3.3.3. Calculation of 2-MIB Degradation Rate

The degradation rate of 2-MIB was calculated according to the following formula:

$$\text{Degradation Rate} = (1 - C_0/C_t) \times 100\%. \quad (1)$$

In Formula (1), C_t represents the residual concentration of 2-MIB (ng/L), and C_0 represents the initial concentration of 2-MIB (ng/L).

The degradation reaction of 2-MIB followed first-order reaction kinetics:

$$\text{Ln}(C_0/C_t) = kt. \quad (2)$$

In Formula (2), t is the reaction time, and k is the reaction rate constant.

4. Conclusions

In this study, the conditions for the laccase-catalyzed degradation of 2-MIB were successfully optimized, achieving a degradation efficiency of approximately 78% under the optimal conditions of pH 4.0, 25 °C, 150 r/min, and 0.1 U/mL after 4 h. Compared with the degradation processes of laccase catalysis and electrochemical oxidation alone, laccase catalysis assisted by a micro-electric field exhibited a synergetic effect for 2-MIB degradation and improved the degradation efficiency to 90.78%. The application of an external micro-electric field may have accelerated electron transfer during oxidation. The degradation pathways of these three catalytic strategies were suggested based on the SPME-GC/MS results. Some degradation intermediate products appeared in all three pathways, and their final degradation products were ketone compounds. This study provides new insights into the degradation of T&O substances catalyzed by laccase, highlighting the innovative approach of combining enzymatic catalysis with electrocatalysis for enhanced degradation efficiency and reduced odor intensity. Future research will focus on developing new methods or optimizing conditions for the complete degradation of 2-MIB.

Supplementary Materials: The following supporting information can be downloaded at: https://www.mdpi.com/article/10.3390/catal14090649/s1, Figure S1: Time profile of 2-MIB degra-dation rate at different pH; Figure S2: Time profile of 2-MIB degradation by laccase at different temperature; Figure S3: Time profile of 2-MIB degradation by laccase at different stirring rate; Figure S4: Time profile of 2-MIB degradation by laccase at different laccase dosage; Figure S5: Time profile of 2-

MIB degradation by laccase at different initial 2-MIB concentration; Figure S6: Lineweaver-Burk plots of free laccase. Table S1: Kinetic parameters of laccase; Table S2: Summary of intermediates determined by SPME-GC/MS in 2-MIB degradation catalyzed by laccase; Table S3: Summary of intermediates determined by SPME-GC/MS in 2-MIB degradation catalyzed by laccase assisted with micro-electric field.

Author Contributions: Conceptualization, L.X. and F.W.; methodology, J.X., S.H. and F.W.; software, T.L. and F.W.; validation, B.L., T.L. and L.X.; formal analysis, L.X. and T.L.; investigation, T.L.; resources, Y.C. and S.H.; data curation, A.M.; writing—original draft preparation, L.X., B.L. and F.W.; writing—review and editing, B.L. and F.W.; visualization, J.Q. and A.M.; supervision, J.X. and F.W.; project administration, G.Z.; funding acquisition, Y.C. and G.Z. All authors have read and agreed to the published version of the manuscript.

Funding: This work was supported by the China Postdoctoral Science Foundation (2022M711379) and Weiqiao-UCAS Special Projects on Low-Carbon Technology Development (GYY-NYHJ-2023-WT-001).

Data Availability Statement: Data are contained within the article and Supplementary Materials.

Acknowledgments: We thank Xianjin Ou of the Institute of Biophysics, Chinese Academy of Sciences, for technical assistance.

Conflicts of Interest: The authors declare no conflicts of interest.

References

1. Li, L.; Yang, S.; Yu, S.; Zhang, Y. Variation and removal of 2-MIB in full-scale treatment plants with source water from Lake Tai, China. *Water Res.* **2019**, *162*, 180–189. [CrossRef] [PubMed]
2. Yu, C.; Shi, C.; Tang, J.; Ji, Q.; Wang, X.; Xu, X.; Wang, G. Release of taste and odour compounds during Zizania latifolia decay: A microcosm system study. *Environ. Pollut.* **2019**, *254*, 112954. [CrossRef]
3. Wang, F.; Li, X.; Liu, T.; Li, X.; Cui, Y.; Xu, L.; Huo, S.; Zou, B.; Qian, J.; Ma, A.; et al. Removal of Taste and Odor Compounds from Water: Methods, Mechanism and Prospects. *Catalysts* **2023**, *13*, 1356. [CrossRef]
4. Zahraei, S.K.; Salemi, A.; Schmidt, T.C. Sample preparation for determination of water taste and odor compounds: A review. *Trends Environ. Anal. Chem.* **2021**, *32*, e00149. [CrossRef]
5. Kim, T.-K.; Moon, B.-R.; Kim, T.; Kim, M.-K.; Zoh, K.-D. Degradation mechanisms of geosmin and 2-MIB during UV photolysis and UV/chlorine reactions. *Chemosphere* **2016**, *162*, 157–164. [CrossRef]
6. Soyluoglu, M.; Kim, D.; Zaker, Y.; Karanfil, T. Removal mechanisms of geosmin and MIB by oxygen nanobubbles during water treatment. *Chem. Eng. J.* **2022**, *443*, 136535. [CrossRef]
7. Abu Hasan, H.; Muhammad, M.H.; Ismail, N.I. A review of biological drinking water treatment technologies for contaminants removal from polluted water resources. *J. Water Process Eng.* **2020**, *33*, 101035. [CrossRef]
8. Mustapha, S.; Tijani, J.O.; Ndamitso, M.M.; Abdulkareem, A.S.; Shuaib, D.T.; Mohammed, A.K. A critical review on geosmin and 2-methylisoborneol in water: Sources, effects, detection, and removal techniques. *Environ. Monit. Assess.* **2021**, *193*, 204. [CrossRef]
9. Barbusinski, K.; Kalemba, K.; Kasperczyk, D.; Urbaniec, K.; Kozik, V. Biological methods for odor treatment—A review. *J. Clean. Prod.* **2017**, *152*, 223–241. [CrossRef]
10. Cococeanu, A.L.; Man, T.E. Methods and Characteristics of Conventional Water Treatment Technologies. In *Water Safety, Security and Sustainability: Threat Detection and Mitigation*; Springer: Cham, Switzerland, 2021; pp. 305–330.
11. Zheng, Z.-X.; Lin, Y.-L.; Fang, R.-F.; Zhou, X.-Y.; Liu, Z.; Dong, Z.-Y.; Zhang, T.-Y.; Xu, B. Removal of algae and algogenic odor compounds via combined pre-chlorination and powdered activated carbon adsorption for source water pretreatment. *Sep. Purif. Technol.* **2023**, *304*, 122365. [CrossRef]
12. Piezer, K.; Li, L.; Jeon, Y.; Kadudula, A.; Seo, Y. The Application of Potassium Permanganate to Treat Cyanobacteria-Laden Water: A Review. *Process Saf. Environ. Prot.* **2021**, *148*, 400–414. [CrossRef]
13. Du, K.; Zhou, B.; Yuan, R. Biodegradation of 2-methylisoborneol by single bacterium in culture media and river water environment. *Int. J. Environ. Stud.* **2017**, *74*, 399–411. [CrossRef]
14. Drummond, L.; von Wallbrunn, C.; Buchhaupt, M. Microbial Degradation of 2-Methylisoborneol in Forest Soil**. *Chem. Biodivers.* **2022**, *19*, e202100734. [CrossRef]
15. Du, H.; Wang, Z.; Sun, Y.; Shah, K.J. An Overview of the Progress made in Research on Odor Removal in Water Treatment Plants. *Water* **2024**, *16*, 280. [CrossRef]
16. Shao, X.; Du, K. Biodegradation of 2-methylisoborneol by enzyme separated from Pseudomonas mandelii. *Water Supply* **2020**, *20*, 2096–2105. [CrossRef]
17. Chmelova, D.; Ondrejovic, M.; Miertus, S. Laccases as Effective Tools in the Removal of Pharmaceutical Products from Aquatic Systems. *Life* **2024**, *14*, 230. [CrossRef]

18. Dong, C.-D.; Tiwari, A.; Anisha, G.S.; Chen, C.-W.; Singh, A.; Haldar, D.; Patel, A.K.; Singhania, R.R. Laccase: A potential biocatalyst for pollutant degradation*. *Environ. Pollut.* **2023**, *319*, 120999. [CrossRef]
19. Bilal, M.; Rasheed, T.; Nabeel, F.; Iqbal, H.M.N.; Zhao, Y. Hazardous contaminants in the environment and their laccase-assisted degradation—A review. *J. Environ. Manag.* **2019**, *234*, 253–264. [CrossRef]
20. Jia, J.; Xue, P.; Ma, L.; Li, P.; Xu, C. Deep degradation of atrazine in water using co-immobilized laccase-1-hydroxybenzotriazole-Pd as composite biocatalyst. *J. Hazard. Mater.* **2024**, *468*, 133779. [CrossRef]
21. Min, Y.; Mei, S.-C.; Pan, X.-Q.; Chen, J.-J.; Yu, H.-Q.; Xiong, Y. Mimicking reductive dehalogenases for efficient electrocatalytic water dechlorination. *Nat. Commun.* **2023**, *14*, 5134. [CrossRef]
22. AlJaberi, F.Y.; Ahmed, S.A.; Makki, H.F.; Naje, A.S.; Zwain, H.M.; Salman, A.D.; Juzsakova, T.; Viktor, S.; Van, B.; Le, P.-C.; et al. Recent advances and applicable flexibility potential of electrochemical processes for wastewater treatment. *Sci. Total Environ.* **2023**, *867*, 161361. [CrossRef] [PubMed]
23. Wei, K.; Shen, C.; Han, W.; Li, J.; Sun, X.; Shen, J.; Wang, L. Advance treatment of chemical industrial tailwater by integrated electrochemical technologies: Electrocatalysis, electrodialysis and electro-microfiltration. *Chem. Eng. J.* **2017**, *310*, 13–21. [CrossRef]
24. Panizza, M.; Cerisola, G. Direct And Mediated Anodic Oxidation of Organic Pollutants. *Chem. Rev.* **2009**, *109*, 6541–6569. [CrossRef] [PubMed]
25. Zhao, C.; Yang, X.; Zhao, B.; Zhang, Z.; Guo, W.; Shen, A.; Ye, M.; Wang, W. Copper porphyrin MOF/graphene oxide composite membrane with high efficiency electrocatalysis and structural stability for wastewater treatment. *J. Membr. Sci.* **2024**, *695*, 122499. [CrossRef]
26. Wang, F.; Owusu-Fordjour, M.; Xu, L.; Ding, Z.; Gu, Z. Immobilization of Laccase on Magnetic Chelator Nanoparticles for Apple Juice Clarification in Magnetically Stabilized Fluidized Bed. *Front. Bioeng. Biotechnol.* **2020**, *8*, 589. [CrossRef]
27. Cao, X.; Gao, J.; Yang, Y.; Li, H.; Zheng, X.; Liu, G.; Jiang, Y. Synergistic degradation of chlorophenol pollutants by a photo-enzyme integrated catalyst. *J. Environ. Chem. Eng.* **2022**, *10*, 107909. [CrossRef]
28. Pang, S.; Wu, Y.; Zhang, X.; Li, B.; Ouyang, J.; Ding, M. Immobilization of laccase via adsorption onto bimodal mesoporous Zr-MOF. *Process Biochem.* **2016**, *51*, 229–239. [CrossRef]
29. Saoudi, O.; Ghaouar, N. Biocatalytic characterization of free and immobilized laccase from Trametes versicolor in its activation zone. *Int. J. Biol. Macromol.* **2019**, *128*, 681–691. [CrossRef]
30. Dai, Y.; Yao, J.; Song, Y.; Liu, X.; Wang, S.; Yuan, Y. Enhanced performance of immobilized laccase in electrospun fibrous membranes by carbon nanotubes modification and its application for bisphenol A removal from water. *J. Hazard. Mater.* **2016**, *317*, 485–493. [CrossRef]
31. Zhou, W.; Zhang, W.; Cai, Y. Laccase immobilization for water purification: A comprehensive review. *Chem. Eng. J.* **2021**, *403*, 126272. [CrossRef]
32. Deng, J.; Wang, H.; Zhan, H.; Wu, C.; Huang, Y.; Yang, B.; Mosa, A.; Ling, W. Catalyzed degradation of polycyclic aromatic hydrocarbons by recoverable magnetic chitosan immobilized laccase from Trametes versicolor. *Chemosphere* **2022**, *301*, 134753. [CrossRef] [PubMed]
33. Schwienheer, C.; Prinz, A.; Zeiner, T.; Merz, J. Separation of active laccases from Pleurotus sapidus culture supernatant using aqueous two-phase systems in centrifugal partition chromatography. *J. Chromatogr. B-Anal. Technol. Biomed. Life Sci.* **2015**, *1002*, 1–7. [CrossRef] [PubMed]
34. Wojtusik, M.; Zurita, M.; Villar, J.C.; Ladero, M.; Garcia-Ochoa, F. Influence of fluid dynamic conditions on enzymatic hydrolysis of lignocellulosic biomass: Effect of mass transfer rate. *Bioresour. Technol.* **2016**, *216*, 28–35. [CrossRef] [PubMed]
35. Sutar, R.S.; Rathod, V.K. Ultrasound assisted Laccase catalyzed degradation of Ciprofloxacin hydrochloride. *J. Ind. Eng. Chem.* **2015**, *31*, 276–282. [CrossRef]
36. Du, J.; Cao, Y.; Liu, G.; Zhao, J.; Li, X.; Qu, Y. Identifying and overcoming the effect of mass transfer limitation on decreased yield in enzymatic hydrolysis of lignocellulose at high solid concentrations. *Bioresour. Technol.* **2017**, *229*, 88–95. [CrossRef] [PubMed]
37. Escalona, I.; de Grooth, J.; Font, J.; Nijmeijer, K. Removal of BPA by enzyme polymerization using NF membranes. *J. Membr. Sci.* **2014**, *468*, 192–201. [CrossRef]
38. Yusof, N. Production of Chitosan Oligosaccharides using β-glycosidic degrading enzyme: Optimization using Response Surface Methodology. *Malays. J. Appl. Sci.* **2020**, *31*, 30–44. [CrossRef]
39. Xue, Q.; Shimizu, K.; Sakharkar, M.K.; Utsumi, M.; Cao, G.; Li, M.; Zhang, Z.; Sugiura, N. Geosmin degradation by seasonal biofilm from a biological treatment facility. *Environ. Sci. Pollut. Res.* **2012**, *19*, 700–707. [CrossRef]
40. Abd El-Hack, M.E.; El-Saadony, M.T.; Elbestawy, A.R.; Ellakany, H.F.; Abaza, S.S.; Geneedy, A.M.; Salem, H.M.; Taha, A.E.; Swelum, A.A.; Omer, F.A.; et al. Undesirable odour substances (geosmin and 2-methylisoborneol) in water environment: Sources, impacts and removal strategies. *Mar. Pollut. Bull.* **2022**, *178*, 113579. [CrossRef]
41. Zhang, K.; Yang, W.; Liu, Y.; Zhang, K.; Chen, Y.; Yin, X. Laccase immobilized on chitosan-coated Fe_3O_4 nanoparticles as reusable biocatalyst for degradation of chlorophenol. *J. Mol. Struct.* **2020**, *1220*, 128769. [CrossRef]
42. Camarero, S.; Ibarra, D.; Martínez, M.J.; Martínez, A.T. Lignin-derived compounds as efficient laccase mediators for decolorization of different types of recalcitrant dyes. *Appl. Environ. Microbiol.* **2005**, *71*, 1775–1784. [CrossRef] [PubMed]
43. Jones, S.M.; Solomon, E.I. Electron transfer and reaction mechanism of laccases. *Cell. Mol. Life Sci.* **2015**, *72*, 869–883. [CrossRef] [PubMed]

44. Wang, C.; Zhang, H.; Ren, D.; Li, Q.; Zhang, S.; Feng, T. Effect of Direct-Current Electric Field on Enzymatic Activity and the Concentration of Laccase. *Indian J. Microbiol.* **2015**, *55*, 278–284. [CrossRef] [PubMed]
45. Zhang, Y.; Jiang, W.; Dong, H.; Hu, X.; Fang, B.; Gao, G.; Zhao, R. Study on the Electrochemical Removal Mechanism of Oxytetracycline by a Ti/IrO_2-Ta_2O_5 Plate. *Int. J. Environ. Res. Public Health* **2021**, *18*, 1708. [CrossRef] [PubMed]
46. Shi, S.; Xu, J.; Zeng, Q.; Liu, J.; Hou, Y.; Jiang, B. Impacts of applied voltage on EMBR treating phenol wastewater: Performance and membrane antifouling mechanism. *Bioresour. Technol.* **2019**, *282*, 56–62. [CrossRef] [PubMed]
47. Yuan, Y.; Guo, S.-H.; Li, F.-M.; Li, T.-T. Effect of an electric field on *n*-hexadecane microbial degradation in contaminated soil. *Int. Biodeterior. Biodegrad.* **2013**, *77*, 78–84. [CrossRef]
48. Li, T.; Guo, S.; Wu, B.; Li, F.; Niu, Z. Effect of electric intensity on the microbial degradation of petroleum pollutants in soil. *J. Environ. Sci.* **2010**, *22*, 1381–1386. [CrossRef]
49. Li, F.; Guo, S.; Wu, B.; Wang, S. Pilot-scale electro-bioremediation of heavily PAH-contaminated soil from an abandoned coking plant site. *Chemosphere* **2020**, *244*, 125467. [CrossRef]
50. Bu, L.; Zhou, S.; Shi, Z.; Deng, L.; Gao, N. Removal of 2-MIB and geosmin by electrogenerated persulfate: Performance, mechanism and pathways. *Chemosphere* **2017**, *168*, 1309–1316. [CrossRef]
51. Wu, H.; Liu, J.; Zhou, X.; Liu, F.; Bai, X.; Wang, R.; Xu, H.; Tan, L.; Zhang, J. Degradation of Bisphenols by Micro-Nano Bubbles Assisted Laccase: Kinetics, Michaelis-Menten Kinetics, Degradation Pathway, and Transformation Relationship. *Catal. Lett.* **2024**, 1–14. [CrossRef]
52. Raj, R.; Tripathi, A.; Das, S.; Ghangrekar, M.M. Removal of caffeine from wastewater using electrochemical advanced oxidation process: A mini review. *Case Stud. Chem. Environ. Eng.* **2021**, *4*, 100129. [CrossRef]
53. Comninellis, C. Electrocatalysis in the electrochemical conversion/combustion of organic pollutants for waste water treatment. *Electrochim. Acta* **1994**, *39*, 1857–1862. [CrossRef]
54. Huo, S.H.; Liu, J.Z.; Zhu, F.F.; Basheer, S.; Necas, D.; Zhang, R.C.; Li, K.; Chen, D.J.; Cheng, P.F.; Cobb, K.; et al. Post treatment of swine anaerobic effluent by weak electric field following intermittent vacuum assisted adjustment of N:P ratio for oil-rich filamentous microalgae production. *Bioresour. Technol.* **2020**, *314*, 123718. [CrossRef] [PubMed]

catalysts

MDPI

Article

High-Performance Photocatalytic Degradation—A ZnO Nanocomposite Co-Doped with Gd: A Systematic Study

Aeshah Alasmari [1], Nadi Mlihan Alresheedi [2], Mohammed A. Alzahrani [3], Fahad M. Aldosari [4], Mostafa Ghasemi [5], Atef Ismail [6] and Abdelaziz M. Aboraia [6,*]

1 Department of Physics, College of Science, University of Bisha, P.O. Box 551, Bisha 61922, Saudi Arabia
2 Department of General Studies, Royal Commission for Jubail and Yanbu, Yanbu Industrial College, P.O. Box 30436, Yanbu 41912, Saudi Arabia
3 Mining Engineering Department, Faculty of Engineering, King Abdulaziz University, P.O. Box 80204, Jeddah 21589, Saudi Arabia
4 Department of Physics, College of Science and Humanities, Prince Sattam Bin Abdulaziz University, P.O. Box 173, Al-Kharj 11942, Saudi Arabia
5 Chemical Engineering Section, Faculty of Engineering, Sohar University, Sohar 311, Oman
6 Department of Physics, Faculty of Science, Al-Azhar University, Assiut 71524, Egypt
* Correspondence: a.m.aboraia@gmail.com

Abstract: This research aims to analyze the improvement in the photocatalytic properties of ZnO nanoparticles by incorporating Gd. In order to understand the influence of incorporating Gd into the ZnO matrix, the photocatalytic activity of the material is compared at various Gd concentrations. Different doping concentrations of Gd ranging from 0 to 0.075 are incorporated into ZnO and the synthesized ZnO-Gd nanocomposites are investigated using structural, morphological, and optical analyses using XRD, SEM, and UV-vis spectroscopy, respectively. The photocatalytic performance of the synthesized ZnO-Gd nanocomposites is determined via the degradation of organic contaminants under visible light. Regarding the latter, the results suggest that photocatalytic efficiency increases with increasing Gd doping levels up to an optimal doping concentration. The enhancement of the photocatalytic performance of Gd-doped ZnO is explained, along with the mechanism related to the availability of new pathways for charge carrier recombination. Among all of them, the 0.075 Gd-doped ZnO catalyst exhibits the highest photocatalytic activity which degrades 89% of MB dye after being irradiated with UV light for 120 min. However, pure ZnO degrades only 40% of MB dye within the same testing conditions. In closing, this work confirms the applicability of Gd-doped ZnO nanocomposites as photocatalysts in cleaning up the environment and in wastewater treatment.

Keywords: ZnO nanoparticles; photocatalysis; Mott–Schottky; band gap; sol–gel

1. Introduction

Polluted water is any water that is harming the environment or has the potential to harm it and the effects of this pollution necessitate immediate action [1–3]. One of the leading causes of this pollution is the release of variable and hazardous organic pollutants, such as organic dyes and pharmaceutical waste effluent, in the wastewater stream [4,5]. These pollutants lead to deforestation, reduced clean and drinkable water, dire environmental consequences, gruesome sicknesses, and even genetic alterations in the long run [6,7]. Most of the conventional physical and chemical treatment methods, such as flocculation, coagulation, reverse osmosis, filtration, etc., are limited in the sense that they cannot remove the persistent dyes in water effectively [8,9]. Innovative solutions are needed to alter these dyes into non-hazardous byproducts, and photocatalysis appears to be a viable option for the remediation of this sort of wastewater [10–12]. Contaminant dyes have been degraded by photocatalytic processes using different semiconductor materials, such as Cu_2O, ZnO, TiO_2, and CuO [13–15]. Of all these materials, zinc oxide (ZnO) is preferred because it is cheap, environmentally friendly, available, and possesses desirable physical and chemical characteristics [16,17].

Zinc oxide nanoparticles (NPs) are easily available wide-band-gap semiconductor materials that have potential applications in electronic, cosmetic, and biomedical industries and environmental remediation [18–21]. Besides their inherent characteristics, such as crystal structure, morphology, and bulk composition, defects also have critical roles in determining any of their properties. In general, defects in ZnO NPs are unfavorable from a crystallography point of view; however, defects can be useful for new applications, such as ferromagnetism, p-type conductivity, spintronics, optoelectronics, etc. [22,23]. In ZnO, various other structure defects such as zinc interstitial (Zni), oxygen interstitial (Oi), zinc vacancy (VZn), and oxygen vacancy (Vo) are also present [24,25]. These defects can change the physical nature of ZnO, for example, by converting diamagnetic ZnO into ferromagnetic, changing its photoluminescence properties, or tailoring the band gap [26,27]. The electron energy levels related to intrinsic defects in ZnO can affect the photoinduced transitions, thus affecting the probability of electron transition photoconductivity and the photocatalytic activity of ZnO NPs [28,29].

Numerous studies have shown that the photocatalytic properties of ZnO can be improved by doping with rare-earth elements (Ce, Eu, Sr, La, Gd, etc.) and transition metals (Mn, Ni, Cr, Cu, etc.), and by forming composites such as ZnO/ZnS, ZnO/SnO_2, ZnO/CuO, ZnO/Fe_2O_3, etc. [30–33]. Young-Hoon Kim et al. [34] and Danilenko et al. [35] have demonstrated the enhanced photocatalytic performance of ZnO by increasing the density of Vo, which is also supported by several publications. As the active sites of ZnO NPs are locally influenced by these defects, the effect of dopant types and doping levels on tuning ZnO's photocatalytic properties remains a challenging research area. Doping with rare earth ions has attracted significant interest due to the electronic transitions occurring within their 4f energy shells. Gd-doped ZnO NPs offer greater advantages and application potential due to their temperature-independent UV and visible luminescence, which is applicable in light-emitting displays, optical storage systems, drug delivery, and catalysis. Magnetic fields and temperature have been found to significantly impact Gd-doped ZnO. Another important property of Gd-doped ZnO is that, in addition to electron conductivity in ZnO, hole conductivity is also enhanced, as holes are more active in the 4f orbital of Gd than electrons. Gd-doped ZnO with varying dopant concentrations ($Gd_xZn_{1-x}O$, where x = 0.00, 0.025, 0.05, and 0.075) was synthesized using the sol–gel combustion method, and its efficiency in degrading MB under UV-light illumination was investigated. The process of doping has significant effects on the size, shape, and photocatalytic performance of ZnO NPs.

2. Result and Discussion

2.1. XRD Analysis

Figure 1 shows the XRD profiles of the synthesized $Zn_{1-x}Gd_xO$ samples (x = 0, 0.025, 0.05, and 0.075). Herein, we identified the hexagonal phase structure for the samples: pure ZnO and ZnO doped with different concentrations of Gd. It is also called the hexagonal phase and is set in the space group of P63mc. This structure has been found to consist of a series of Zn–O columns with tetrahedral coordination that repeat along a c-axis. For ZnO doped with Gd, the phase structure investigated will be hexagonal, as with pure ZnO, but with some Zn ions replaced by Gd ions. This has been categorized as a substitutional doping mechanism. When Gd ions are incorporated into the ZnO lattice, the lattice strain and defects such as vacancies or interstitials can be generated, altering the physical and chemical properties of the material. Rietveld refinement is a simple technique of studying the crystal structures of materials, such as ZnO and ZnO-Gd. In this case, the program models the crystal structure, compares it with the X-ray diffraction pattern observed, and calculates the changes in lattice parameters, atomic positions, and thermal vibrations. Rietveld refinement of the XRD data allows estimates of the phase purity, crystallite size, and lattice strain of the material needed to assess the impact of doping on the crystal structure to be obtained. Consequently, the structure properties of the synthesized pure ZnO and ZnO doped with Gd are expected to have a hexagonal phase structure, and, based on this structure, the Rietveld refinement analysis can be employed to compare the crystal structure. The incorporation of rare earth ions (Gd^{3+}/Gd^{2+}) into the ZnO lattice resulted in a subtle shift towards lower 2θ angles in the XRD pattern, observed primarily in the dominant (101) diffraction peak. The larger size of introduced rare earth ions compared to Zn^{2+} acts like a crowbar, stretching the ZnO crystal lattice (unit cell expansion) and disrupting its perfect order (reduced crystallinity). This effect manifests as broader peaks in the X-ray data, similar to findings by Suneel Kumar et al. [34].

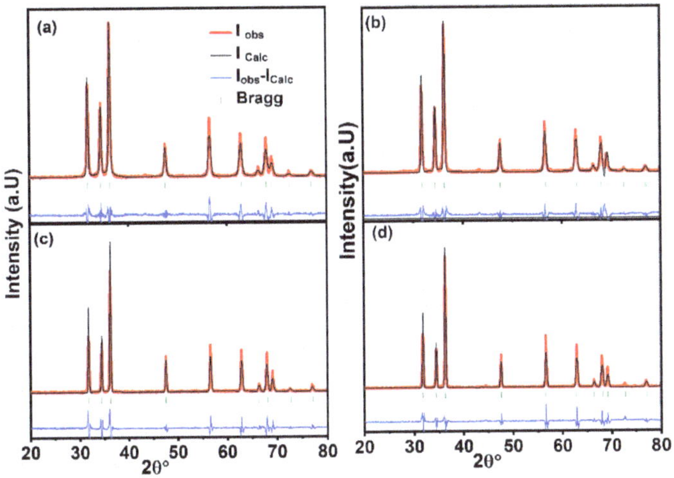

Figure 1. XRD patterns of (**a**) ZnO, (**b**) 0.025 Gd-doped ZnO, (**c**) 0.05 Gd-doped ZnO, and (**d**) 0.075 Gd-doped ZnO.

2.2. FTIR Analysis

The FTIR spectroscopy route was utilized to verify the existence of functional groups in both pristine and ZnO doped with Gd, as depicted in Figure 2a,b. The FTIR peak observed at around 443 cm^{-1} in pure ZnO is assigned to the lattice metal–oxygen bond (Zn-O), as shown in Figure 2b. Upon the analysis of FTIR, the peak detected at around 2924 cm^{-1} is associated with symmetric and asymmetric stretching vibrations of CH_2 adsorbed on the surface [36]. The widened peak observed in the range of 3100–3658 cm^{-1} signifies the stretching vibration of the O-H hydroxyl group due to surface-adsorbed water. In contrast to pristine nanoparticles, the ZnO doped with 5% Gd displays a broadening of the peak at the 417 cm^{-1} band position, indicating lattice distortion at the bond of Zn-O due to the presence of Gd dopant ions. Moreover, the introduction of Gd doping causes a shift in the FTIR spectrum towards higher wave numbers, signifying changes in bond length resulting from the substitution of Zn through Gd^{3+} ions. There are two extra peaks detected in the spectrum at 1388 and 1519 cm^{-1}, confirming the presence of Gd.

2.3. UV-Vis Spectrum

The optical characteristics of the nanoparticles produced were examined using UV-vis spectroscopy. The spectrum of the UV absorption of both the pristine and Gd-doped ZnO preparations is depicted in Figure 3a. Analysis of the UV-vis spectrum indicates that the intensity of the absorption cut-off peak diminishes with higher concentrations of Gd, indicative of reduced crystallinity. Pure ZnO exhibits an absorption wavelength at around 94 nm, corresponding to a band gap of 3.15 eV. In comparison, 2.5, 5, and 7.5% Gd-doped ZnO display a redshift, with cut-off wavelengths around 400 and 403 nm and energy gaps of 3.1 eV and 3.07 eV, correspondingly. The decrease in the band gap of Gd-doped ZnO is ascribed to the formation of narrow (medium) energy levels involving the ZnO band gap energy levels [14]. The shift towards longer wavelengths in the absorption peak of 2.5, 5, and 7.5% Gd-doped ZnO is ascribed to alterations in the defect structure within the lattice crystal and the formation of oxygen vacancies within the ZnO system of the crystal. However, ZnO doped with 0.05 Gd exhibits a blue shift relative to pristine zinc oxide through an absorption peak at around 390 nm and an energy gap of around 3.17 eV, as shown in Figure 3b. The augmented energy gap observed in ZnO doped with 0.05 Gd suggests a reduction in crystallite size compared to pristine zinc oxide [16]. The findings suggest that the heightened band gap in ZnO doped with 0.05 Gd is a consequence of the quantum confinement effect. Typically, as the crystallite size rises, the direct energy gap energy of nanostructured materials also grows, and noticeable changes in the morphology of the nanoparticles are observed in comparison to pure ZnO.

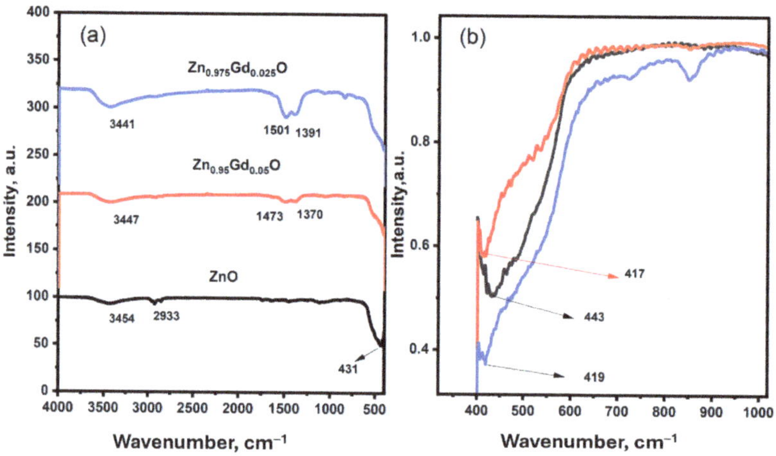

Figure 2. FTIR spectroscopy of (**a**) ZnO, 0.05 Gd-doped ZnO, and 0.075 Gd-doped ZnO and (**b**) selected peaks at around 417, 419, 443 cm^{-1} for ZnO, 0.05 Gd-doped ZnO, and 0.075 Gd-doped ZnO.

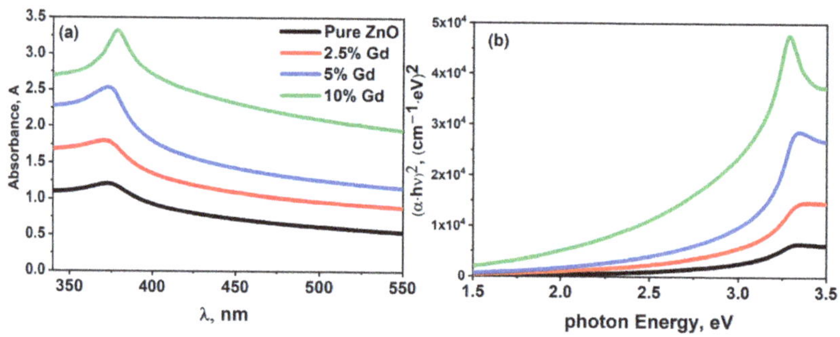

Figure 3. UV-vis spectroscopy (**a**,**b**) energy gap calculated of ZnO, 0.025 Gd-doped ZnO, 0.05 Gd-doped ZnO, and 0.075 Gd-doped ZnO.

Variations in the nanoparticle energy gap energy depend on various factors such as preparation conditions, shape, grain size, and alterations in lattice parameters within the crystal structure. While bulk ZnO possesses a direct energy gap of around 3.14 eV, in nanostructured zinc oxide, electronic states become discrete at the level of individual molecules and atoms. Consequently, the red shift of the energy gap and electronic transitions is restricted when electronic populations occupy discrete states, leading to a notable blue shift with an increase in the energy gap. The observed increase in the band gap of ZnO doped with 0.05 Gd is primarily attributed to the quantum confinement effect, where the forbidden bandwidth enlarges as crystal size decreases. At the nanoscale, ZnO crystals exhibit high densities in the UV region and maximum transparency in the visible region due to electronic transitions from O2p to Zn3d orbitals. The introduction of Gd into ZnO alters the optical band gap, largely due to surface defects or oxygen vacancies within the ZnO crystal system. These vacancies or defects decrease crystal size, effectively widening the forbidden band gap. UV studies reveal that ZnO doped with 0.05 Gd exhibits increased photon absorption efficiency compared to pristine ZnO during photocatalysis.

2.4. Morphological Properties

Figure 4 demonstrates the scanning electron microscopy (SEM) photos of synthesized $Zn_{1-x}Gd_xO$ (x = 0, 0.025, 0.05, and 0.075) samples. SEM is a helpful method that can be used to investigate the morphology and microstructure of materials, in particular $Zn_{1-x}Gd_xO$ (x = 0, 0.025, 0.05, and 0.075). These characterizations using SEM give information on the particle size, shape, and distribution of the material, which may be altered by the doping concentration of Gd. Using SEM images, the as-prepared pure ZnO (x = 0) consists of hexagonal platelets or rods that have smooth surfaces and almost similar sizes. The platelets or rods are generally of the order of several microns in length and a few hundred nanometers in breadth; they adhere remarkably closely together and are arranged randomly. A high percentage of Gd added into the ZnO lattice brings about a change in the morphology and microstructure of the material. At low doping concentrations (x = 0.025, 0.05) in the SEM images, the dopant may elicit hexagonal platelet or rod-like morphology similar to that of ZnO, but with a smaller particle size and distorted forms. The particle size is anticipated to decrease with increasing Gd concentration because the lattice strain and defects created by the dopant ions lead to a reduction in particle size. The surface of the particles may also change toward being rougher and/or more porous, reflecting the development of defects and/or secondary phases. However, at higher doping concentrations (x = 0.075), SEM images may possess a higher degree of structural complexity with the presence of agglomerates, clusters, or secondary phases. This will probably coalesce into larger and more random-shaped particles, and there may be less distinct hexagonal platelet or rod-like shapes. These X-ray diffractions may be due to the limited solubility of Gd in ZnO, suggesting the possible existence of Gd-rich regions or compounds.

Figure 4. SEM photos of (**a**) ZnO, (**b**) 0.025 Gd-doped ZnO, (**c**) 0.05 Gd-doped ZnO, (**d**) 0.075 Gd-doped ZnO.

2.5. Photocatalytic Activity of $Zn_{1-x}Gd_xO$ (x = 0, 0.025, 0.05 and 0.075)

The photodegradation of MB under UV light was investigated for the following $Zn_{1-x}GdxO$ samples: ZnO, $Zn_{0.975}Gd_{0.025}O$, $Zn_{0.95}Gd_{0.05}O$, and $Zn_{0.925}Gd_{0.075.}$ In a normal experiment, 0.025 g of zinc oxide, gadolinium oxide, or $Zn_{1-x}Gd_xO$ (x = 0, 0.025, 0.05, and 0.075) was mixed with 10 ppm of MB dye solution with a volume of 50 mL. Finally, the sample was filtered and centrifuged for 10 min at 5000 r.p.m, after which the sample was measured using UV-vis spectroscopy. Then, 30 min dark pre-illumination was carried out to achieve adsorption equilibrium between $Zn_{1-x}Gd_xO$ (where x = 0, 0.025, 0.05, and 0.075) photocatalyst and MB solution. These MB molecules, which were adsorbed onto the $Zn_{1-x}Gd_xO$ surface (with x = 0, 0.025, 0.05, and 0.075), interacted with the electron–hole pairs activated by UV light. A closed dark reactor with a 10 cm gap between the UV lamp and the MB-$Zn_{1-x}Gd_xO$ (where x = 0, 0.025, 0.05, and 0.075) beaker was used. Consequently, the percent photodegradation of the MB solution was determined by measuring its absorbance at 665 nm. The photodegradation/removal efficiency was calculated using the following equation:

$$Removal\% = \frac{\left(C_i - C_f\right)}{C_i} \times 100$$

Following the completion of the photocatalytic degradation process, the starting and final MB concentrations (C_i and C_f) were calculated.

2.6. Photocatalytic Activity and Mechanism of the Reaction

We investigated the photocatalytic degradation of methylene blue (MB) dye in ZnO suspension using the photo-Fenton reaction and compared its efficiency with ZnO-doped Gd prepared through the co-precipitation route. ZnO is a promising candidate for photocatalysts used in practical and large-scale applications, as it can be produced at a low cost from non-toxic materials exhibiting a high level of activity. Nevertheless, the photocatalytic ability of ZnO can be enhanced by doping it with rare-earth metal similar to Gd. For this investigation, ZnO pristine and 0.025, 0.05, and 0.075 Gd ZnO photocatalysts were based on a sol–gel method. These catalysts were tested by degrading MB dye under UV light irradiation. Figure 5a–d shows the photocatalytic performance of pure and Gd-doped ZnO catalysts.

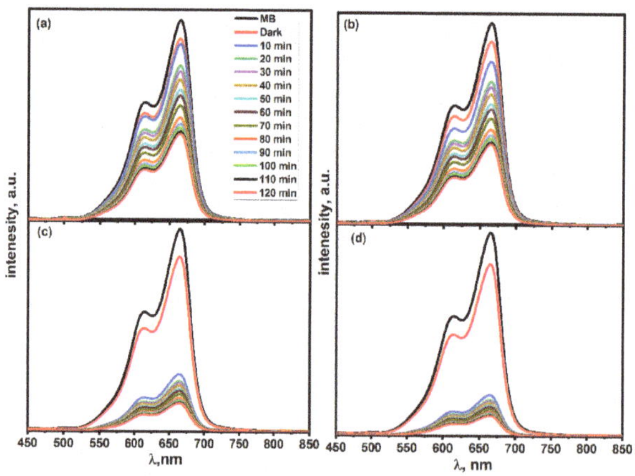

Figure 5. Photocatalytic degradation of (**a**) ZnO, (**b**) 0.025 Gd-doped ZnO, (**c**) 0.05 Gd-doped ZnO, and (**d**) 0.075 Gd-doped ZnO.

Among all, 0.075 Gd-doped ZnO catalyst exhibited the highest photocatalytic activity which degraded 89% of MB dye after being irradiated with UV light for 120 min, as shown in Figure 6. However, pure ZnO degraded only 40% of MB dye within the same testing conditions. The increase in the photocatalytic activity of Gd-doped ZnO catalysts is associated with a defect band caused by introduced gadolinium ions, which serve as electron traps that can inhibit charge recombination [11]. To assess the photocatalytic efficiency of a catalyst, the kinetic rate constant turned out to be a useful parameter. In this study, a series of experiments for the degradation of methylene blue (MB) dye using pure ZnO and different Gd-doped ZnO photocatalysts (0, 0.025, 0.05, and 0.075 wt%) were investigated using the Langmuir–Hinshelwood kinetic model to calculate the kinetic rate constants. The kinetic rate constant for pure ZnO was determined to be 0.0065 min^{-1} and that for 0.025 Gd-doped ZnO was found to be 0.0068 min^{-1}. Nonetheless, the kinetic rate constant was found to rise with increasing Gd doping concentration, up to its maximum of 0.00723 min^{-1} using 0.075 Gd-doped ZnO, as demonstrated in Figure 7. As mentioned earlier, the increase in the kinetic rate constant with Gd doping may be explained by the better photocatalytic activity of Gd-doped ZnO. The Gd ions play the role of electron trap centers and the inhibition of electron–hole recombination leads to the formation of ROS. ROS are involved in the degradation of MB dye; the higher the concentration of ROS, the higher the rate of degradation. The kinetic rate constant of 0.075 Gd-doped ZnO-NPs is more than that of pure ZnO which shows that this catalyst has high efficiency in degrading the MB dye. The increase in the rate constant for kinetics also confirms that 0.075 Gd-doped ZnO possesses superior photocatalytic activity due to the inserted Gd ions, as demonstrated in Table 1.

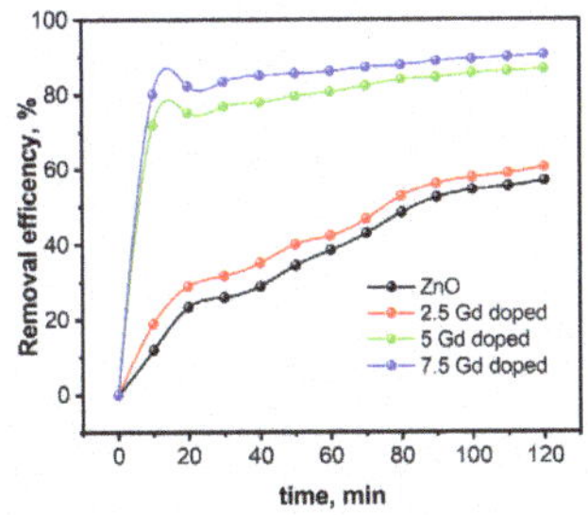

Figure 6. Removal efficiency of ZnO, 0.025 Gd-doped ZnO, 0.05 Gd-doped ZnO, and 0.075 Gd-doped ZnO.

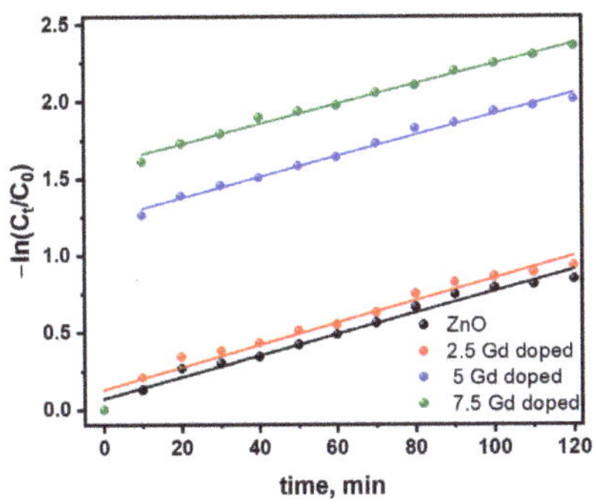

Figure 7. –ln C_t/C_0 vs. time of ZnO, 0.025 Gd-doped ZnO, 0.05 Gd-doped ZnO, and 0.075 Gd-doped ZnO.

Impedance spectroscopy is a potent means for investigating the electrical characteristics of materials, such as semiconductors and photocatalysts. In this work, the impedance of pure ZnO, as well as 0, 0.025, 0.05, and 0.075 Gd-doped ZnO nanoparticles, was recorded at room temperature as a function of frequency. The impedance spectra were fitted with the equivalent circuit model, and the charge transfer resistance (Rct) was obtained. Pure ZnO has a higher impedance, which suggests a lower charge transfer rate along with high electron–hole pair recombination. On the other hand, it was observed that with an increased concentration of Gd doping, higher impedance values are obtained, which indicates an increase in the rate of charge transfer and a decrease in the probability of recombination between electron–hole pairs. The highest resistance against charge transfer between electrode and material is referred to as charge transfer resistance (Rct). The impedance spectra were analyzed using the equivalent circuit model to extract Rct values of pure ZnO, as well as 0, 0.025, 0.05, and 0.075 Gd-doped ZnO nanoparticles, as shown in Figure 8. The high Rct value of pristine ZnO reflects a low charge transfer rate, together with the large recombination process of electron–hole pairs. In contrast, the Rct value decreases remarkably with rising Gd doping levels, which can be attributed to faster charge transfer and lower electron–hole pair recombination, as exhibited in Figure 8.

Table 1. Rate constant values of ZnO, 0.025 Gd-doped ZnO, 0.05 Gd-doped ZnO, and 0.075 Gd-doped ZnO.

Samples	Rate Constant	R^2
ZnO	0.00654	98
2.5% Gd-doped ZnO	0.00683	96
5% Gd-doped ZnO	0.00696	99
7.5% Gd-doped ZnO	0.00723	99

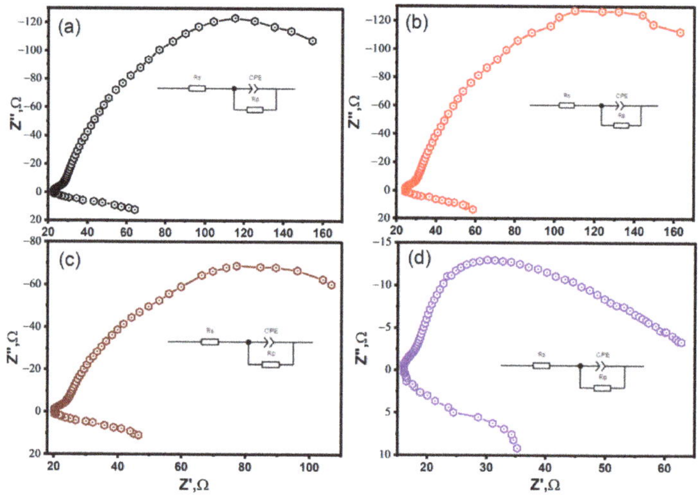

Figure 8. EIS of (**a**) ZnO (in set equivalent circuit), (**b**) 0.025 Gd-doped ZnO (in set equivalent circuit), (**c**) 0.05 Gd-doped ZnO (in set equivalent circuit), and (**d**) 0.075 Gd-doped ZnO (in set equivalent circuit).

Cole–Cole analysis, which is missing in previous studies but was carried out in this study, shows that the Rct value decreases during the spectral analysis of the impedance semicircle and increases in the Gd doping concentration, therefore reducing the rate of recombination of electron–hole pairs and improving charge transfer. It can be inferred from these results that the electrical properties of ZnO nanoparticles can be improved through Gd doping to enhance its photocatalytic activity. Mott–Schottky analysis is a comparative technique largely employed in the investigations of electric semiconductor materials, including zinc oxide and its nanoparticles, as shown in Figure 9.

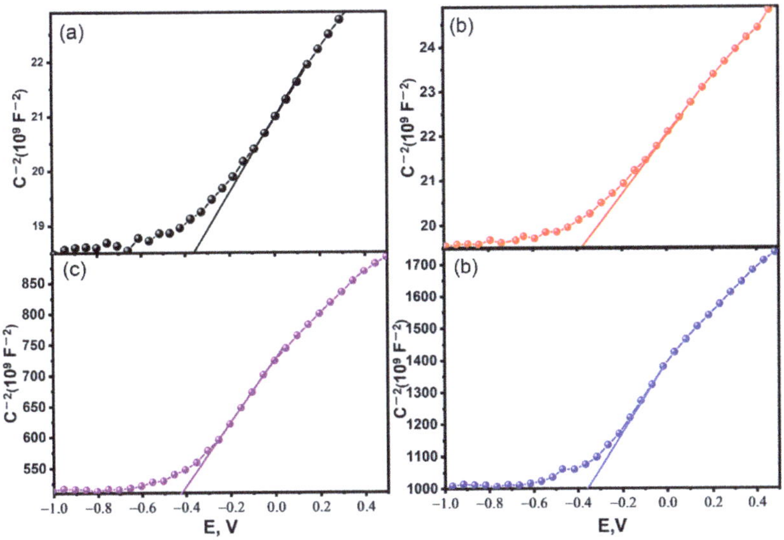

Figure 9. Mott-Schotky plot of (**a**) ZnO, (**b**) 0.025 Gd-doped ZnO, (**c**) 0.05 Gd-doped ZnO, and (**d**) 0.075 Gd-doped ZnO.

The Mott–Schottky analysis of pure ZnO and 0.025, 0.05, and 0.075 Gd-doped ZnO nanoparticles was carried out in order to find their flat-band potential (Vfb) and donor density (Nd) values. The Mott–Schottky plot is a graphical representation of the capacitance of the space charge region as a function of the applied voltage. In this work, Mott–Schottky plots were obtained by applying an oscillating voltage to the sample and measuring the capacitance at a frequency of 500 Hz. The Mott–Schottky plot depicts the variations of Nd while the intercept with the plot's vertical axis depicts the Nfb. The Mott-Schottky plots obtained for 0, 0.025, 0.05, and 0.075 Gd-doped ZnO show a positive slope, indicating the expansion of p-type semiconductors. The values of the flat-band potential (Vfb) and density of donor ions (Nd) were calculated from the graphs of the Mott–Schottky plots. For pure ZnO, Vfb = −0.31 V/Ag/Agcl; however, Vfb drops as the Gd doping concentration increases.

The flat-band potential (Vfb) for undoped ZnO was determined to be −0.31 V versus Ag/AgCl; in contrast, the value of flat-band potential (Vfb) was found to increase with an increase in the concentration of Gd dopant. The flat-band potential (Vfb) values of 0.025 Gd/doped ZnO, 0.05 Gd/doped ZnO, and 0.075 Gd-doped samples were −0.33, −0.37, and −0.30 V, respectively. The increase in Vfb with an increase in Gd concentration can be attributed to the addition of Gd ions, which introduce a distorted structure in the lattice of ZnO and modify the electronic structure. The enhancement of donor density (Nd) with Gd concentration also supports the fact that Gd ions replace some ions in the lattice of ZnO, thus creating an electronic state within the band gap. This finding indicates that Gd dopants exhibit potential for improving the electrical conductivity of ZnO nanoparticles and their effectiveness as photocatalysts.

2.7. Photocatalyst Growth Mechanism

This proposed degradation mechanism for Gd co-doped ZnO is described in Figure 10 below. This is true when light with a wavelength equal to or greater than the bandgap energy of ZnO interacts with the material when electrons and holes are created. By doping the Gd, the generated electrons are effectively annihilated, hence reducing the recombination rate. These trapped electrons are then able to react with oxygen, forming superoxide radicals. At the same time, hydroxyl ions interact with holes forming hydroxyl radicals. The generated OH• and O2•$^-$ radicals that are formed play a role in breaking the dye molecules. Gd co-doped ZnO exhibits a higher degradation rate compared to pure ZnO due to two primary factors: decreased recombination rate and the enhanced generation of OH• and O2•$^-$ radicals.

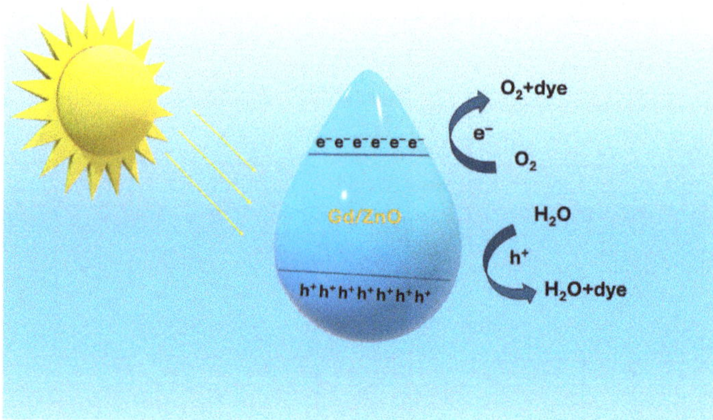

Figure 10. Photocatalytic reaction mechanism.

To assess the stability of the synthesized photocatalysts, a recycling experiment was conducted under UV-vis light irradiation, and the results are presented in Figure 11. Five consecutive cycles of MB dye degradation were performed using Gd/ZnO nanoparticles. After each cycle, the photocatalyst was recovered through ultracentrifugation, washed with acetone and deionized water, and dried at 60 °C for 1 h. To compensate for the approximately 17% weight loss during the recycling process, a small amount of fresh photocatalyst was added to the recovered sample for each subsequent cycle. The recycling experiment demonstrates the high stability of Gd/ZnO nanoparticles, making them a promising candidate for organic pollutant degradation.

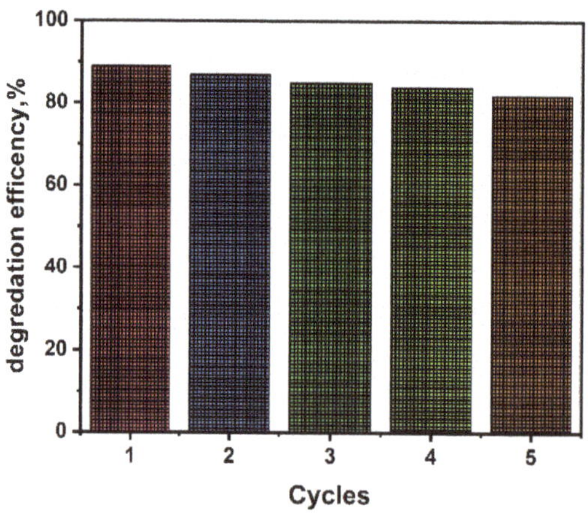

Figure 11. Recycle test of 0.075 Gd-doped ZnO.

3. Experimental Work

3.1. $Zn_{1-x}Gd_xO$ Synthesis

All the required chemicals for our research study were identified by Loba Company, including zinc nitrate, citric acetate, samarium oxide, and isopropanol. As part of our research study, $Zn_{1-x}Gd_xO$ (x = 0, 0.025, 0.05, and 0.075) nanoparticles were synthesized using the sol–gel combustion method. In this typical procedure, a precursor solution containing zinc nitrated, gadolinium oxide, and citric acid dissolved into isopropanol (50 mL) was prepared in a beaker. The citric acid solution was also stirred at room temperature until homogeny without losing the stoichiometric ratios of citric acid to metal cation. The mixture was then heated on a magnetic stirrer up to approximately 80 °C and stirred until it transformed into a gel. When heated further to the ignition point, the material burst into flames and resulted in large lightweight ash. The obtained material was heat-treated to a temperature of around 600 °C and then milled to a very fine particle size. The gel combustion technique has benefits over other methods because of its easiness and cheapness, in addition to high rates of yield. Moreover, this approach aligns well with the principles of the circular economy, emphasizing effectiveness and sustainability.

3.2. Characterization

Phase- and micro-characterization studies of $Zn_{1-x}Gd_xO$ (x = 0, 0.025, 0.05, and 0.075) nanoparticles were conducted using XRD, FTIR, and SEM analyses. FTIR analysis was carried out using a BRUKER Tensor 37 FTIR spectrophotometer, with a scan at wave number intervals of 4 m^{-1} to 40 m^{-1}. Electron microscopy was performed on the Jeol JFC 1100 E at a scale of 20,000 times and an accelerating voltage of 25 KV. The XRD patterns were obtained with the LAN Scientific diffractometer equipped with a monochromatized Cu-kα radiation source operating at 30 kV and 16 mA. The optical parameters of the synthesized $Zn_{1-x}Gd_xO$ were analyzed using a UV–visible spectrophotometer (JASCO, V 570) in the range of 200 to 1000 nm. The impedance and Mott–Schottky tests were studied using workstation Cortest Cs305. The photocatalysis experiment was conducted with a xenon lamp of 500 watts, with 25 mg of photocatalyst materials.

4. Conclusions

$Zn_{1-x}Gd_xO$ (0, 0.025, 0.05, and 0.075) was synthesized via the sol–gel combustion route. The XRD proved that all materials have the same structural phase (hexagonal phase) and do not appear as a new additional phase from other impurities. UV-vis spectrum analysis indicates that the intensity of the absorption cut-off peak diminishes with higher concentrations of Gd, indicative of reduced crystallinity. Pure ZnO exhibits an absorption wavelength at around ~394 nm, corresponding to a band gap of 3.15 eV. In comparison, 0.025, 0.05, and 0.075 Gd-doped ZnO display a redshift, with cut-off wavelengths around 400 and 403 nm and energy gaps of 3.1 eV and 3.07 eV, correspondingly. The decrease in the band gap of Gd-doped ZnO is ascribed to the formation of narrow (medium) energy levels involving the ZnO band gap energy levels. Using SEM images, the as-prepared pure ZnO (x = 0) consists of hexagonal platelets or rods that have smooth surfaces and almost similar sizes. The platelets or rods are generally of the order of several microns in length and a few hundred nanometers in breadth; they adhere remarkably closely together and are arranged randomly. A high percentage of Gd added into the ZnO lattice brings about a change in the morphology and microstructure of the material. Among them all, the 0.075 Gd-doped ZnO catalyst exhibits the highest photocatalytic activity, which degrades 89% of MB dye after being irradiated with UV light for 120 min. However, pure ZnO degrades only 40% of MB dye within the same testing conditions. It was observed that with increased concentrations of Gd doping, higher impedance values are obtained, which indicates an increase in the rate of charge transfer and a decrease in the probability of recombination between electron–hole pairs.

Author Contributions: Conceptualization, A.A. and N.M.A.; methodology, M.A.A.; software, F.M.A.; validation, M.G.; formal analysis, A.I.; investigation, A.M.A. All authors have read and agreed to the published version of the manuscript.

Funding: The authors are thankful to the Deanship of Graduate Studies and Scientific Research at the University of Bisha for supporting this work through the Fast-Track Research Support Program.

Data Availability Statement: The original contributions presented in the study are included in the article; further inquiries can be directed to the corresponding author.

Conflicts of Interest: The authors declare no conflicts of interest.

References

1. Aboraia, A.M.; Al-omoush, M.; Solayman, M.; Saad, H.M.H.; Khabiri, G.; Saad, M.; Alsulaim, G.M.; Soldatov, A.V.; Ismail, Y.A.M.; Gomaa, H. A heterostructural MoS2QDs@UiO-66 nanocomposite for the highly efficient photocatalytic degradation of methylene blue under visible light and simulated sunlight. *RSC Adv.* **2023**, *13*, 34598–34609. [CrossRef] [PubMed]
2. Guo, X.; Zhang, F.; Zhang, Y.; Hu, J. Review on the advancement of SnS 2 in photocatalysis. *J. Mater. Chem. A* **2023**, *11*, 7331–7343. [CrossRef]
3. Ashraf, S.; Sandhu, Z.A.; Raza, M.A.; Bhalli, A.H.; Hamayun, M.; Ashraf, A.; Al-Sehemi, A.G. Unveiling Bi-functional potential of ZnMoO4-enriched nanoflakes modified electrodes for efficient photocatalysis and supercapacitors. *J. Sol-Gel Sci. Technol.* **2024**, *112*, 25–43. [CrossRef]
4. Rathi, B.S.; Kumar, P.S.; Vo, D.-V.N. Critical review on hazardous pollutants in water environment: Occurrence, monitoring, fate, removal technologies and risk assessment. *Sci. Total Environ.* **2021**, *797*, 149134. [CrossRef]
5. Alsukaibi, A.K.D. Various approaches for the detoxification of toxic dyes in wastewater. *Processes* **2022**, *10*, 1968. [CrossRef]
6. Gaur, N.; Narasimhulu, K.; PydiSetty, Y. Recent advances in the bio-remediation of persistent organic pollutants and its effect on environment. *J. Clean. Prod.* **2018**, *198*, 1602–1631. [CrossRef]
7. Behera, M.; Nayak, J.; Banerjee, S.; Chakrabortty, S.; Tripathy, S.K. A review on the treatment of textile industry waste effluents towards the development of efficient mitigation strategy: An integrated system design approach. *J. Environ. Chem. Eng.* **2021**, *9*, 105277. [CrossRef]
8. Saravanan, A.; Kumar, P.S.; Jeevanantham, S.; Karishma, S.; Tajsabreen, B.; Yaashikaa, P.R.; Reshma, B. Effective water/wastewater treatment methodologies for toxic pollutants removal: Processes and applications towards sustainable development. *Chemosphere* **2021**, *280*, 130595. [CrossRef] [PubMed]
9. Khan, M.D.; Singh, A.; Khan, M.Z.; Tabraiz, S.; Sheikh, J. Current perspectives, recent advancements, and efficiencies of various dye-containing wastewater treatment technologies. *J. Water Process Eng.* **2023**, *53*, 103579. [CrossRef]
10. Ahmed, M.; Mavukkandy, M.O.; Giwa, A.; Elektorowicz, M.; Katsou, E.; Khelifi, O.; Naddeo, V.; Hasan, S.W. Recent developments in hazardous pollutants removal from wastewater and water reuse within a circular economy. *NPJ Clean Water* **2022**, *5*, 1–25. [CrossRef]
11. Hynes, N.R.J.; Kumar, J.S.; Kamyab, H.; Sujana, J.A.J.; Al-Khashman, O.A.; Kuslu, Y.; Ene, A.; Kumar, B.S. Modern enabling techniques and adsorbents based dye removal with sustainability concerns in textile industrial sector-A comprehensive review. *J. Clean. Prod.* **2020**, *272*, 122636. [CrossRef]
12. Kumari, H.; Sonia; Suman; Ranga, R.; Chahal, S.; Devi, S.; Sharma, S.; Kumar, S.; Kumar, P.; Kumar, S. A review on photocatalysis used for wastewater treatment: Dye degradation. *Water Air Soil Pollut.* **2023**, *234*, 349. [CrossRef]
13. Ajmal, A.; Majeed, I.; Malik, R.N.; Iqbal, M.; Nadeem, M.A.; Hussain, I.; Yousaf, S.; Mustafa, G.; Zafar, M.I.; Nadeem, M.A. Photocatalytic degradation of textile dyes on Cu2O-CuO/TiO_2 anatase powders. *J. Environ. Chem. Eng.* **2016**, *4*, 2138–2146. [CrossRef]
14. Mosleh, S.; Rahimi, M.R.; Ghaedi, M.; Dashtian, K.; Hajati, S. Sonochemical-assisted synthesis of CuO/Cu_2O/Cu nanoparticles as efficient photocatalyst for simultaneous degradation of pollutant dyes in rotating packed bed reactor: LED illumination and central composite design optimization. *Ultrason. Sonochem.* **2018**, *40*, 601–610. [CrossRef]
15. Uma, B.; Anantharaju, K.S.; Renuka, L.; Nagabhushana, H.; Malini, S.; More, S.S.; Vidya, Y.S.; Meena, S. Controlled synthesis of (CuO-Cu_2O) Cu/ZnO multi oxide nanocomposites by facile combustion route: A potential photocatalytic, antimicrobial and anticancer activity. *Ceram. Int.* **2021**, *47*, 14829–14844. [CrossRef]
16. Elhosiny Ali, H.; Ganesh, V.; Haritha, L.; Aboraia, A.M.; Hegazy, H.H.; Butova, V.; Soldatov, A.V.; Algarni, H.; Guda, A.; Zahran, H.Y.; et al. Kramers-Kronig analysis of the optical linearity and nonlinearity of nanostructured Ga-doped ZnO thin films. *Opt. Laser Technol.* **2021**, *135*, 106691. [CrossRef]
17. Alharshan, G.A.; Aboraia, A.M.; Uosif, M.A.M.; Sharaf, I.M.; Shaaban, E.R.; Saad, M.; Almohiy, H.; Elsenety, M.M. Optical band gap tuning, DFT understandings, and photocatalysis performance of ZnO nanoparticle-doped Fe compounds. *Materials* **2023**, *16*, 2676. [CrossRef] [PubMed]
18. Goswami, S.; Bishnoi, A.; Tank, D.; Patel, P.; Chahar, M.; Khaturia, S.; Modi, N.; Khalid, M.; Alam, M.W.; Yadav, V.K. Recent trends in the synthesis, characterization and commercial applications of zinc oxide nanoparticles—A review. *Inorganica Chim. Acta* **2024**, *573*, 122350. [CrossRef]
19. Terna, A.D.; Elemike, E.E.; Mbonu, J.I.; Osafile, O.E.; Ezeani, R.O. The future of semiconductors nanoparticles: Synthesis, properties and applications. *Mater. Sci. Eng. B* **2021**, *272*, 115363. [CrossRef]
20. Gaur, J.; Kumar, S.; Pal, M.; Kaur, H.; Batoo, K.M.; Momoh, J.O. Current trends: Zinc oxide nanoparticles preparation via chemical and green method for the photocatalytic degradation of various organic dyes. *Hybrid Adv.* **2023**, *5*, 100128. [CrossRef]

21. Irede, E.L.; Awoyemi, R.F.; Owolabi, B.; Aworinde, O.R.; Kajola, R.O.; Hazeez, A.; Raji, A.A.; Ganiyu, L.O.; Onukwuli, C.O.; Onivefu, A.P. Cutting-edge developments in zinc oxide nanoparticles: Synthesis and applications for enhanced antimicrobial and UV protection in healthcare solutions. *RSC Adv.* **2024**, *14*, 20992–21034. [CrossRef] [PubMed]
22. Nunes, D.; Pimentel, A.; Santos, L.; Barquinha, P.; Pereira, L.; Fortunato, E.; Martins, R. *Metal Oxide Nanostructures: Synthesis, Properties and Applications*; Elsevier: Amsterdam, The Netherlands, 2018.
23. Jayadevan, K.P.; Tseng, T.-Y. One-dimensional ZnO nanostructures. *J. Nanosci. Nanotechnol.* **2012**, *12*, 4409–4457. [CrossRef] [PubMed]
24. Vidya, R.; Ravindran, P.; Fjellvåg, H.; Svensson, B.G.; Monakhov, E.; Ganchenkova, M.; Nieminen, R.M. Energetics of intrinsic defects and their complexes in ZnO investigated by density functional calculations. *Phys. Rev. B—Condens. Matter Mater. Phys.* **2011**, *83*, 045206. [CrossRef]
25. Wang, Z.; Dai, W.; Fu, X. The Role of the Defects on the Photocatalytic Reactions on ZnO. In *UV-Visible Photocatalysis for Clean Energy Production and Pollution Remediation: Materials, Reaction Mechanisms, and Applications*; John Wiley & Sons: Hoboken, NJ, USA, 2023; pp. 59–75.
26. Sharma, D.K.; Shukla, S.; Sharma, K.K.; Kumar, V. A review on ZnO: Fundamental properties and applications. *Mater. Today Proc.* **2022**, *49*, 3028–3035. [CrossRef]
27. Punia, K.; Lal, G.; Dolia, S.N.; Kumar, S. Defects and oxygen vacancies tailored structural, optical, photoluminescence and magnetic properties of Li doped ZnO nanohexagons. *Ceram. Int.* **2020**, *46*, 12296–12317. [CrossRef]
28. Nguyen, X.-S.; Nguyen, M.-Q.; Trinh, X.-T.; Joita, A.C.; Nistor, S.V. Correlation of native point defects and photocatalytic activity of annealed ZnO nanoparticle studied by electron spin resonance and photoluminescence emission. *Semicond. Sci. Technol.* **2020**, *35*, 095035. [CrossRef]
29. Li, Y.; Li, Y.; Yin, Y.; Xia, D.; Ding, H.; Ding, C.; Wu, J.; Yan, Y.; Liu, Y.; Chen, N. Facile synthesis of highly efficient ZnO/ZnFe2O4 photocatalyst using earth-abundant sphalerite and its visible light photocatalytic activity. *Appl. Catal. B Environ.* **2018**, *226*, 324–336. [CrossRef]
30. Nami, M.; Sheibani, S.; Rashchi, F. Photocatalytic performance of coupled semiconductor ZnO–CuO nanocomposite coating prepared by a facile brass anodization process. *Mater. Sci. Semicond. Process.* **2021**, *135*, 106083. [CrossRef]
31. Chanu, L.A.; Singh, W.J.; Singh, K.J.; Devi, K.N. Effect of operational parameters on the photocatalytic degradation of Methylene blue dye solution using manganese doped ZnO nanoparticles. *Results Phys.* **2019**, *12*, 1230–1237. [CrossRef]
32. Nami, M.; Rakhsha, A.; Sheibani, S.; Abdizadeh, H. The enhanced photocatalytic activity of ZnO nanorods/CuO nanourchins composite prepared by chemical bath precipitation. *Mater. Sci. Eng. B* **2021**, *271*, 115262. [CrossRef]
33. Zare, A.; Saadati, A.; Sheibani, S. Modification of a Z-scheme ZnO-CuO nanocomposite by Ag loading as a highly efficient visible light photocatalyst. *Mater. Res. Bull.* **2023**, *158*, 112048. [CrossRef]
34. Kim, Y.-H.; Kim, S.; Kim, K.-j.; Kim, C.; Jang, J.H.; Kim, Y.-M.; Lee, H. Multiscale probing of the influence of the defect-induced variation of oxygen vacancies on the photocatalytic activity of doped ZnO nanoparticles. *J. Mater. Chem. A* **2020**, *8*, 25345–25354. [CrossRef]
35. Danilenko, I.; Gorban, O.; Maksimchuk, P.; Viagin, O.; Malyukin, Y.; Gorban, S.; Volkova, G.; Glasunova, V.; Mendez-Medrano, M.G.; Colbeau-Justin, C. Photocatalytic activity of ZnO nanopowders: The role of production techniques in the formation of structural defects. *Catal. Today* **2019**, *328*, 99–104. [CrossRef]
36. Jabri, S.; Souissi, H.; Sallet, V.; Lusson, A.; Meftah, A.; Galtier, P.; Oueslati, M. Adsorbed Molecules and Surface Treatment Effect on Optical Properties of ZnO Nanowires Grown by MOCVD. *J. Electron. Mater.* **2017**, *46*, 4690–4694. [CrossRef]

catalysts

MDPI

Article

Sunlight-Driven Synthesis of $TiO_2/(MA)_2SnCl_4$ Nanocomposite Films for Enhanced Photocatalytic Degradation of Organic Pollutants

Abdellah Kaiba [1], Amani M. Alansi [2], Ali Oubelkacem [3], Ilyas Chabri [3], Salah T. Hameed [4], Naveed Afzal [5], Mohsin Rafique [5] and Talal F. Qahtan [1,*]

1 Department of Physics, College of Science and Humanities in Al-Kharj, Prince Sattam bin Abdulaziz University, Al-Kharj 11942, Saudi Arabia; a.kaiba@psau.edu.sa
2 Chemistry Department, College of Science, Taiz University, Taiz 12372, Yemen; mn.alansi2016@gmail.com
3 Laboratory of Materials Physics and Systems Modelling, (LP2MS), Physics Department, Faculty of Science, University Moulay Ismail, Meknes 50050, Morocco; a.oubelkacem@umi.ac.ma (A.O.); ilyasmath9@gmail.com (I.C.)
4 Physics Department, Faculty of Education, Taiz University, Taiz 12372, Yemen; salah1phy@gmail.com
5 Centre for Advanced Studies in Physics, GC University, Lahore 54000, Pakistan; naveedafzal@gcu.edu.pk (N.A.); mohsinrafique@gcu.edu.pk (M.R.)
* Correspondence: t.qahtan@psau.edu.sa or tfqahtan@gmail.com

Abstract: In this study, a $TiO_2/(MA)_2SnCl_4$ nanocomposite film was synthesized using a sustainable, sunlight-driven approach, demonstrating enhanced photocatalytic performance for environmental remediation. TiO_2 nanoparticles (TiO_2-NPs) were dispersed in ethanol and mixed with a methylammonium (MA) and $SnCl_2$ precursor solution, followed by drop-casting onto a glass substrate and exposure to direct sunlight for 2 h. Sunlight served as an energy source, facilitating in situ structural modifications and leading to the formation of a well-integrated $TiO_2/(MA)_2SnCl_4$ hybrid structure, where TiO_2 was effectively encapsulated. Characterization revealed a band gap reduction from 3.1 eV (TiO_2-NPs) to 2.6 eV in the nanocomposite, extending light absorption into the visible range. The formation of Sn–O–Ti interactions enhanced charge separation, minimized electron–hole recombination, and improved charge carrier dynamics. Photocatalytic degradation tests using methylene blue (MB) under sunlight showed that the nanocomposite film achieved 90% MB degradation within 60 min, outperforming TiO_2-NPs, which achieved only 75% degradation. The presence of oxygen vacancies (OVs) generated during sunlight exposure further enhanced photocatalytic efficiency by acting as charge traps and reaction sites. This study introduces a green synthesis strategy leveraging sunlight as a renewable energy source, marking the first integration of $(MA)_2SnCl_4$ with TiO_2-NPs for enhanced photocatalysis. The synergistic effects of extended visible-light absorption, defect engineering, and efficient charge separation make $TiO_2/(MA)_2SnCl_4$ nanocomposite films a scalable, cost-effective solution for water purification applications, offering a promising solar-driven approach to addressing global water contamination challenges.

Keywords: photocatalysis; $TiO_2/(MA)_2SnCl_4$ nanocomposite film; sunlight-driven synthesis; visible light absorption; environmental remediation

1. Introduction

The growing global demand for clean water has sparked extensive research into sustainable water purification technologies. Among these, photocatalysis has emerged as a promising solution due to its ability to utilize sunlight to drive chemical reactions that degrade pollutants in water. Photocatalysis relies on photocatalyst materials that absorb light and generate reactive species capable of breaking down harmful contaminants. This process is not only effective but also environmentally friendly, leveraging solar energy, an abundant and renewable resource. By converting solar energy into chemical energy, photocatalysis offers a cost-effective and sustainable approach to water decontamination, making it a standout method for addressing water pollution on a large scale [1–5].

Titanium dioxide (TiO_2) is one of the most extensively researched photocatalysts, praised for its excellent oxidation capabilities, chemical stability, non-toxicity, and cost-effectiveness [6,7]. Despite these advantages, TiO_2 has a significant drawback, namely its relatively large band gap of around 3.2 eV (in the anatase form), which restricts its light absorption to the ultraviolet (UV) range, accounting for only about 5% of the solar spectrum [8,9]. As a result, the photocatalytic activity of TiO_2 under natural sunlight is limited, making it less effective for practical environmental applications. This narrow UV absorption hampers its ability to harness the full potential of solar energy, thus reducing its overall efficiency for photocatalytic processes.

To address the limitations of TiO_2, numerous strategies have been explored, including doping TiO_2 with metal or nonmetal ions, generating oxygen vacancies (OVs), and coupling TiO_2 with visible-light-responsive materials [10–13]. These modifications aim to extend the absorption of TiO_2 from the UV region into the visible range, thereby enhancing its photocatalytic performance under sunlight. Metal halide perovskites have emerged as promising candidates to improve photocatalytic efficiency due to their tunable band gaps, exceptional light absorption capabilities, and excellent charge carrier mobility [14,15]. Well-studied examples include methylammonium lead iodide ($CH_3NH_3PbI_3$) [16,17] and cesium lead bromide ($CsPbBr_3$) [18]. Despite their remarkable optoelectronic properties, their use in practical applications is limited due to concerns related to lead toxicity and chemical instability [19]. Methylammonium tin chloride ($(MA)_2SnCl_4$) is a non-toxic alternative to conventional lead-based perovskites, mitigating concerns about lead contamination. It offers multiple benefits, such as tunable band gaps, excellent visible light absorption, and high charge carrier mobility. Additionally, its ability to form a heterojunction with other semiconductors enhances photocatalytic efficiency by promoting effective charge separation and reducing recombination rates. These qualities make $(MA)_2SnCl_4$ an ideal material for significantly boosting the photocatalytic performance of TiO_2 [20,21].

There are different approaches to synthesizing TiO_2-based nanocomposites, including sol–gel methods, hydrothermal synthesis, chemical vapor deposition, and impregnation techniques [22,23]. However, many of these methods involve complex setups, high energy consumption, and the use of hazardous chemicals, which can hinder scalability and sustainability. This study employs a green, sunlight-driven synthesis method to create $TiO_2/(MA)_2SnCl_4$ nanocomposite films, offering a scalable and eco-friendly method for photocatalyst production without the need for complex, energy-intensive processes. Utilizing sunlight as the energy source, this in situ synthesis method aligns with green chemistry principles, presenting an effective, sustainable alternative for large-scale photocatalyst development.

The synthesis of the $TiO_2/(MA)_2SnCl_4$ nanocomposite involves dispersing TiO_2 nanoparticles (TiO_2-NPs) in ethanol and mixing them with a precursor solution containing methylammonium (MA) and $SnCl_2$. The resulting mixture is drop-cast onto a glass substrate and exposed to direct sunlight for 2 h, enabling the in situ formation of $(MA)_2SnCl_4$, which effectively encapsulates TiO_2-NPs. This hybrid structure combines the extended light absorption capabilities of Sn- and MA-incorporated phases with the efficient electron transport properties of TiO_2-NPs, enhancing photocatalytic degradation of pollutants under sunlight. Additionally, the immobilization of the photocatalyst on a solid substrate offers ease of reuse, making it a practical solution for scalable environmental applications.

Photocatalytic tests using methylene blue (MB) degradation under sunlight demonstrated that the nanocomposite film achieved 90% degradation within 60 min, significantly surpassing the 75% degradation observed in TiO_2-NPs. This improvement is attributed to band gap reduction (2.6 eV vs. 3.1 eV for TiO_2-NPs), enhanced charge separation, and the presence of OVs that facilitate charge trapping and reactive species generation. The novelty of this study lies in the successful integration of Sn and MA into TiO_2-NPs through a simple, sunlight-driven synthesis approach, utilizing renewable solar energy to drive the formation and interaction of $(MA)_2SnCl_4$ with TiO_2. This cost-effective, scalable synthesis method provides a sustainable pathway for solar-driven water purification, demonstrating the potential of $TiO_2/(MA)_2SnCl_4$ nanocomposite films for practical environmental remediation applications.

2. Results and Discussion

Figure 1 presents Raman spectroscopy and SEM analysis of TiO_2-NPs and $TiO_2/(MA)_2SnCl_4$ nanocomposite films, highlighting the structural and morphological modifications induced by Sn and MA incorporation, as well as the integration of the $(MA)_2SnCl_4$ phase into the TiO_2-NPs framework. Raman spectra (Figure 1a,b) reveal that TiO_2-NPs exhibit distinct anatase peaks at 144, 197, 396, 515, and 640 cm^{-1}, confirming their high crystallinity [24]. In contrast, $TiO_2/(MA)_2SnCl_4$ shows broadened peaks with reduced intensity, suggesting Sn substitution within the TiO_2 lattice, TiO_2 encapsulation, and structural reorganization induced by $(MA)_2SnCl_4$ incorporation.

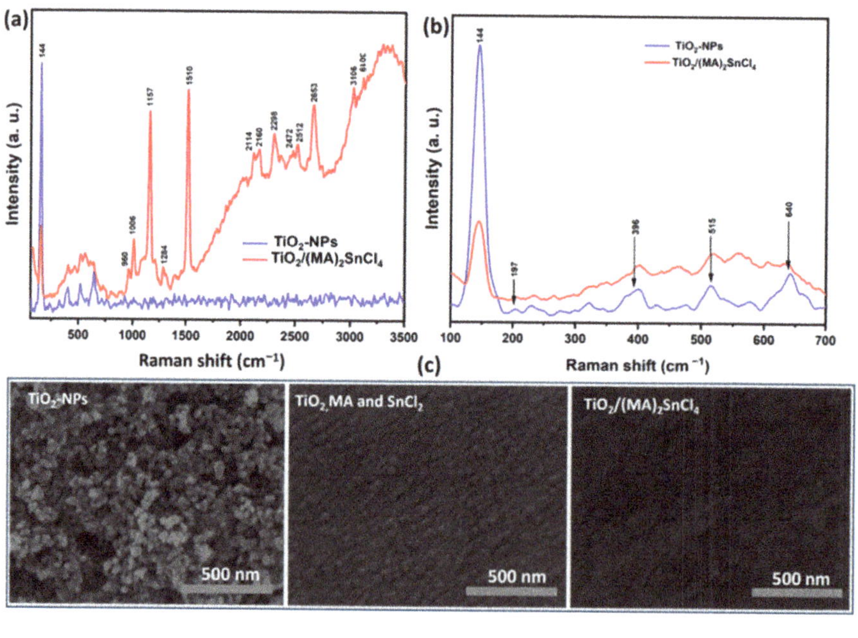

Figure 1. (**a**) Raman spectra of TiO_2-NPs and $TiO_2/(MA)_2SnCl_4$ nanocomposite. (**b**) Magnified Raman spectra in the low-frequency region (100–700 cm^{-1}), highlighting the weakening and broadening of char-acteristic TiO_2 peaks upon interaction with Sn and MA. (**c**) SEM images of TiO_2-NPs, TiO_2 mixed with MA and $SnCl_2$, and $TiO_2/(MA)_2SnCl_4$, illustrating the morphological transformation from agglomerated TiO_2 nanoparticles to a smooth and compact nanocomposite film, confirming en-capsulation and structural reorganization.

The appearance of additional peaks in the 1000–3500 cm^{-1} range, corresponding to C-H, N-H, and CH_3 vibrational modes from MA [25], confirms the successful integration of methylammonium into the hybrid structure, further validating the structural impact of $(MA)_2SnCl_4$ on TiO_2-NPs. The modifications in the low-frequency Raman region (144–640 cm^{-1}) suggest strong interactions between Sn and TiO_2, leading to lattice distortions, defect formation, and heterostructure evolution. These structural transformations, along with weakened TiO_2 vibrational modes, indicate that Sn incorporation and MA stabilization influence the electronic structure and bonding environment, potentially enhancing charge carrier mobility and interfacial charge transfer dynamics. The SEM images support these findings, revealing a transition from agglomerated TiO_2 nanoparticles to a smoother, well-integrated nanocomposite film, confirming the formation of a refined hybrid structure with improved material homogeneity and interfacial interactions.

The SEM images (Figure 1c) illustrate the morphological transformation from agglomerated TiO_2-NPs to a smooth and compact $TiO_2/(MA)_2SnCl_4$ nanocomposite film, confirming structural modifications induced by Sn and MA incorporation. The TiO_2-NPs exhibit a rough, highly agglomerated morphology with loosely packed particles and visible voids, corresponding to the sharp anatase Raman peaks (144, 197, 396, 515, and 640 cm^{-1}), which indicate high crystallinity. As MA and $SnCl_2$ are introduced, the middle-stage sample displays reduced agglomeration and enhanced particle connectivity, signifying initial interactions between TiO_2, MA, and Sn, which are reflected in Raman peak broadening in the 144–640 cm^{-1} range, indicative of lattice distortions and TiO_2 encapsulation effects.

The final $TiO_2/(MA)_2SnCl_4$ nanocomposite exhibits a uniform, dense, and well-organized film-like structure, confirming successful TiO_2 encapsulation within the $(MA)_2SnCl_4$ matrix, as supported by weakened TiO_2 Raman peaks and the emergence of new organic vibrational modes in the 1000–3500 cm^{-1} region. These morphological refinements indicate improved material homogeneity, enhanced interfacial interactions, and better charge transport dynamics, making the nanocomposite an ideal candidate for photocatalytic and optoelectronic applications. The combined findings from SEM and Raman spectroscopy validate that Sn incorporation and MA stabilization lead to a structurally refined hybrid system with superior functional properties.

The XPS data in Figure 2 provides crucial insights into the elemental composition, chemical states, and interactions in the $TiO_2/(MA)_2SnCl_4$ nanocomposite film compared to the TiO_2-NPs film. The survey scan (Figure 2a) confirms the successful incorporation of Sn, Cl, C, N, and O in the nanocomposite, indicating the effective integration of the $(MA)_2SnCl_4$ phase into the TiO_2-NPs matrix. In contrast, the TiO_2-NPs spectrum primarily shows Ti and O, with a small C 1s peak attributed to surface contamination. The additional peaks for Sn 4d, Cl 2p, N 1s, Sn 3d, and Sn 3p in the nanocomposite confirm that Sn, MA, and Cl have effectively interacted with TiO_2, forming a stable hybrid structure even after sunlight exposure.

Figure 2b presents the high-resolution XPS spectra for the N 1s, Cl 2p, and Sn 3d regions, further verifying the integration of $(MA)_2SnCl_4$ into the TiO_2 framework. The N 1s spectrum exhibits three distinct peaks: N1 (398.2 eV), corresponding to nitrogen in a reduced environment (likely associated with MA^+) [26,27], N2 (398.8 eV), attributed to nitrogen involved in hydrogen bonding or interaction with TiO_2 surface groups; and N3 (401.5 eV), linked to nitrogen in a more oxidized environment. The Cl 2p spectrum shows two characteristic peaks, Cl $2p_{3/2}$ (201.7 eV) and Cl $2_{p1/2}$ (199.9 eV), confirming the presence of chloride ions from the $(MA)_2SnCl_4$ phase [28]. The Sn 3d spectrum deconvolutes into two pairs of peaks, corresponding to Sn^{2+} (485.4 eV for Sn $3d_{5/2}$ and 494.0 eV for Sn $3d_{3/2}$) and Sn^{4+} (486.8 eV for Sn $3d_{5/2}$ and 495.7 eV for Sn $3d_{3/2}$), confirming the coexistence of multiple oxidation states [29]. This suggests partial oxidation of Sn^{2+} to

Sn^{4+} during synthesis, likely facilitated by sunlight exposure and atmospheric oxygen, which could impact the material's optoelectronic properties. The combined evidence from the N 1s, Cl 2p, and Sn 3d spectra strongly supports the formation of an $(MA)_2SnCl_4$-based hybrid material, effectively encapsulating the TiO_2-NPs and modifying its surface chemistry. These modifications are expected to enhance interfacial charge transfer and improve the nanocomposite's stability and functional performance in photocatalytic and optoelectronic applications.

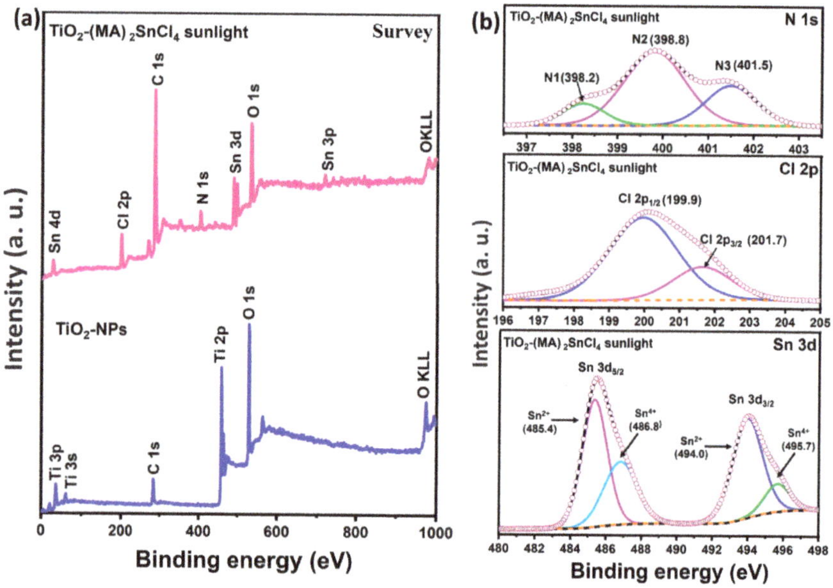

Figure 2. XPS analysis of TiO_2-NPs film and $TiO_2/(MA)_2SnCl_4$ film after sunlight exposure. (**a**) Survey spectra showing Ti, O, C, Sn, Cl, and N. (**b**) High-resolution XPS spectra of N 1s, Cl 2p, and Sn 3d, confirming chemical states and bonding in $TiO_2/(MA)_2SnCl_4$ nanocomposite post-exposure.

The XPS data in Figure 3 compares the C 1s, Ti 2p, and O 1s core-level spectra for TiO_2-NPs and $TiO_2/(MA)_2SnCl_4$ nanocomposite films after sunlight exposure, providing insights into the chemical bonding, surface interactions, and modifications induced by incorporating the $(MA)_2SnCl_4$ phase into TiO_2-NPs. As shown in Figure 3a, both TiO_2-NPs and the nanocomposite exhibit three main C 1s peaks [30,31]: C1 (284.5 eV), attributed to adventitious carbon from surface contamination, corresponding to C–C or C–H bonds; C2 (286.0 eV), associated with C–O or C–OH bonds, likely resulting from atmospheric exposure; and C3 (288.3 eV), representing higher oxidation states of carbon (C=O or O=C–O bonds), possibly from adsorbed CO_2. In the $TiO_2/(MA)_2SnCl_4$ nanocomposite, the C 1s spectrum exhibits a shift to lower binding energies and the emergence of additional carbon environments, confirming the presence of the methylammonium (MA) group. The C1 peak shifts slightly to 284.0 eV, corresponding to C–C or C–H bonds, likely originating from the organic MA group. A new C2 peak at 285.6 eV, attributed to C–N bonds, confirms the incorporation of the MA cation into the nanocomposite [32,33]. The C3 peak remains at 288.3 eV, reflecting oxidized carbon species (C=O or O=C–O), which may arise from surface oxidation or partial degradation under sunlight exposure. These spectral modifications suggest that MA integration significantly alters the nanocomposite's surface chemistry, enhancing interfacial interactions and modifying the electronic environment, which may contribute to improved photocatalytic and optoelectronic properties.

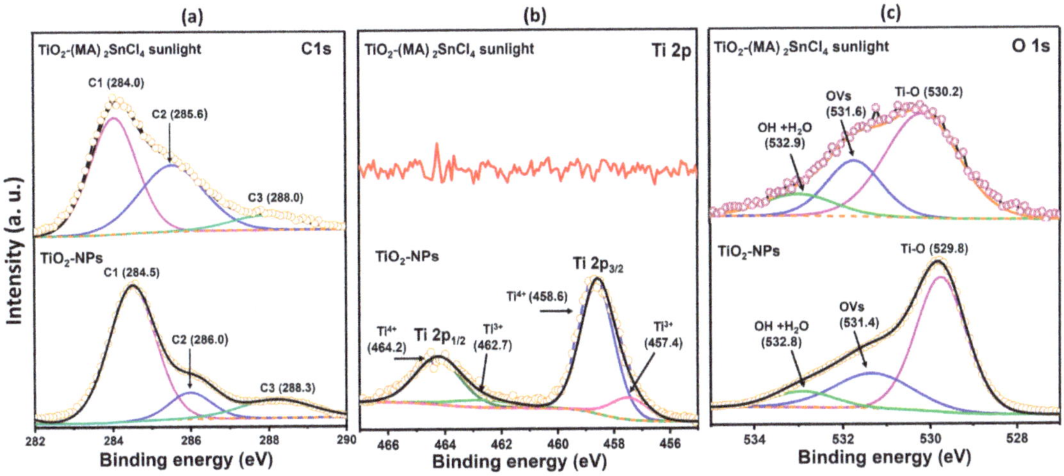

Figure 3. High-resolution XPS spectra of (**a**) C 1s, (**b**) Ti 2p, and (**c**) O 1s for TiO_2-NPs film and $TiO_2/(MA)_2SnCl_4$ film after sunlight exposure, showing changes in carbon bonding, titanium oxidation states, and OVs in samples.

In the TiO_2-NPs (Figure 3b), the Ti 2p spectrum exhibits two distinct peaks for Ti $2p_3/_2$, with a dominant peak at 458.6 eV corresponding to Ti^{4+} in the anatase phase, and a secondary peak at 457.4 eV, attributed to Ti_{3+}, indicating the presence of intrinsic lattice defects associated with oxygen vacancies (OVs), which are known to enhance photocatalytic activity. Additional peaks for Ti $2_{p1}/_2$ at 464.2 eV (Ti^{4+}) further confirm the presence of titanium in its oxidized state, while the peak at 462.7 eV (Ti_{3+}) signifies titanium in a reduced state, reinforcing the existence of OVs and defect-induced electronic states in TiO_2 [34]. In contrast, the Ti 2p signal is notably absent in the $TiO_2/(MA)_2SnCl_4$ nanocomposite, indicating complete encapsulation of TiO_2 by the $(MA)_2SnCl_4$ phase. This encapsulation effectively masks the TiO_2 core from detection, supporting the formation of a uniform hybrid coating that integrates the nanocomposite's components and influences its surface chemistry [35,36].

The O 1s spectrum (Figure 3c) provides further insights into the oxygen environment of both TiO_2-NPs and $TiO_2/(MA)_2SnCl_4$ nanocomposite after sunlight exposure. In TiO_2-NPs, the peak at 529.8 eV corresponds to lattice oxygen (Ti–O bonds) in the anatase crystal structure [37], eflecting a highly ordered lattice with minimal defects. However, in the nanocomposite, this peak shifts slightly to 530.2 eV with reduced intensity, suggesting lattice reorganization due to $(MA)_2SnCl_4$ integration. A peak at 531.4 eV in TiO_2-NPs, associated with defective oxygen sites due to intrinsic lattice defects [38,39], shifts to 531.6 eV in the nanocomposite, with a significant increase in intensity, confirming a higher concentration of OVs. This increase in OVs can be attributed to the dual oxidation states of Sn (Sn^{2+} and Sn^{4+}), which play a critical role in defect formation.

The presence of Sn^{2+} and Sn $^{4+}$ in the nanocomposite promotes oxygen vacancy generation through a charge compensation mechanism. Sn^{2+}, with a lower oxidation state than Ti^{4+}, induces local charge imbalance when substituting Ti sites, leading to oxygen vacancy formation to maintain charge neutrality. Additionally, the partial oxidation of Sn^{2+} to Sn^{4+} during synthesis or sunlight exposure causes the release of lattice oxygen, further increasing the concentration of OVs. These vacancies play a crucial role in facilitating charge separation, reducing electron–hole recombination, and improving overall electronic conductivity, contributing to enhanced functionality. The peak at ~532.8 eV, corresponding to surface hydroxyl groups or adsorbed water, is intensified and broadened in the nanocomposite, indicating an increase in hydroxyl species and adsorbed moisture, which enhances surface reactivity and catalytic performance [40].

The notable rise in OVs and surface defects in the $TiO_2/(MA)_2SnCl_4$ nanocomposite after sunlight exposure suggests structural modifications that contribute to improved charge transport, interfacial charge transfer, and electronic interactions [41]. These observations highlight the profound structural and electronic changes induced by Sn^{2+} and Sn^{4+} incorporation, supporting the formation of a refined hybrid system with enhanced functional properties.

The synthesis of the $TiO_2/(MA)_2SnCl_4$ nanocomposite film (Figure 4) involves a series of chemical interactions and sunlight-driven transformations, ensuring the successful integration of Sn, MA, and TiO_2 into a hybrid structure. Initially, TiO_2 nanoparticles (TiO_2-NPs) are dispersed in ethanol to create a stable colloidal suspension, where ethanol prevents agglomeration by forming hydrogen bonds with surface hydroxyl groups on TiO_2, promoting uniform dispersion. This suspension is then mixed with a precursor solution of methylammonium chloride (MA) and $SnCl_2$, facilitating initial interactions between Sn^{2+} ions and TiO_2 surfaces. The mixture is drop-cast onto a clean glass substrate and dried under ambient conditions, allowing ethanol evaporation, which concentrates MA^+ and Sn ions around TiO_2-NPs, promoting further interactions and nucleation. The dry film is then exposed to direct sunlight for 2 h, utilizing solar energy to drive photochemical reactions and oxidation processes.

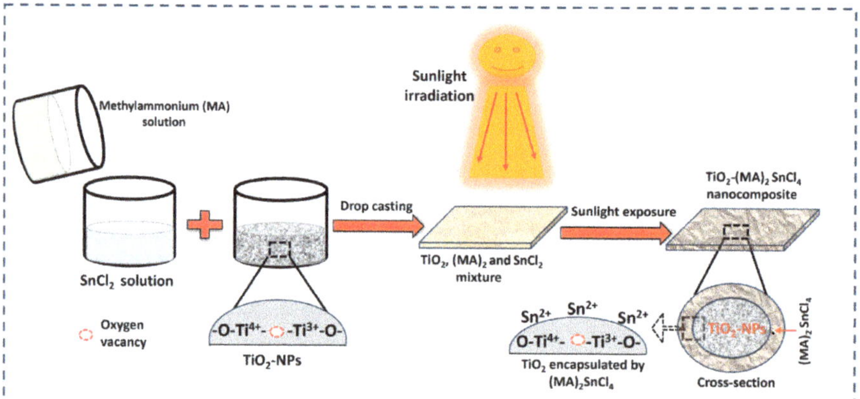

Figure 4. A schematic illustration of the synthesis mechanism for the TiO_2-$(MA)_2SnCl_4$ nanocomposite film under sunlight exposure, highlighting the encapsulation of TiO_2-NPs by the $(MA)_2SnCl_4$.

During the synthesis process, $SnCl_2$ undergoes oxidation and hydrolysis, where Sn^{2+} is partially oxidized to Sn^{4+} due to interactions with photogenerated charge carriers in TiO_2, with holes driving the oxidation reaction. Atmospheric oxygen further facilitates Sn^{2+} to Sn^{4+} conversion, leading to an increased Sn^{2+}/Sn^{4+} ratio, which enhances OVs formation [42]. The coexistence of Sn^{2+} and Sn^{4+} generates OVs in TiO_2, which serve as active sites for charge transport and electronic interactions. In the nanocomposite, Sn–O bonds dominate, forming strong covalent or ionic linkages between Sn^{4+} and oxygen atoms on TiO_2, including lattice oxygen and hydroxyl groups, ensuring structural stability and efficient charge transfer, while MA^+ interacts through hydrogen bonding and electrostatic forces.

Unlike a simple surface modification, TiO_2 is embedded within the composite, forming a well-integrated hybrid structure. This Sn–O-Ti interaction optimizes charge carrier dynamics, facilitating electron transfer from $(MA)_2SnCl_4$ to TiO_2 while holes migrate in the opposite direction, reducing charge recombination. The presence of OVs stabilizes this heterostructure, enhancing charge mobility and improving photocatalytic efficiency. This synergistic interaction between TiO_2, Sn species, and MA significantly enhances light absorption and charge separation, making the nanocomposite highly effective for photoactive applications [43].

Figure 5 presents the optical properties of TiO_2-NPs and $TiO_2/(MA)_2SnCl_4$ nanocomposite films, comparing their absorbance spectra and Tauc plots for band gap estimation. Figure 5a shows that the TiO_2-NPs film exhibits an absorption edge at approximately 386 nm [30], corresponding to a band gap energy of ~3.1 eV, which is characteristic of UV-light-active TiO_2-NPs. In contrast, the $TiO_2/(MA)_2SnCl_4$ nanocomposite demonstrates a red-shifted absorption edge, extending into the visible region up to ~480 nm. This enhanced visible-light absorption results from the incorporation of Sn and MA, which introduce intermediate electronic states within the band gap, modifying the optical transition pathways [44]. The Tauc plots (Figure 5b) reveal a band gap reduction from 3.1 eV in TiO_2-NPs to approximately 2.6 eV in the nanocomposite, attributed to defect engineering and structural modifications.

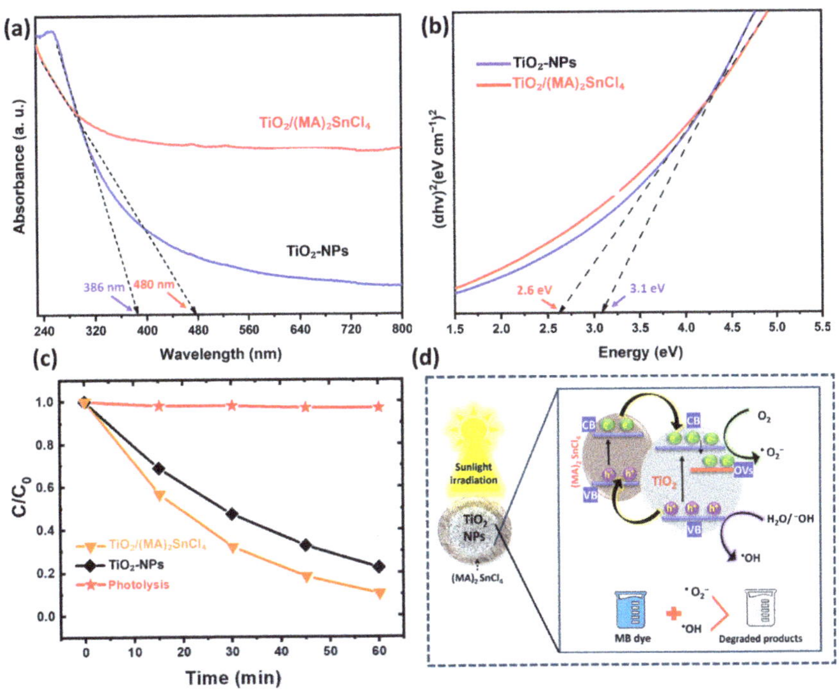

Figure 5. An overview of the optical properties and photocatalytic performance of TiO_2-NPs film and TiO_2-$(MA)_2SnCl_4$ nanocomposite film. (**a**) UV-Vis absorbance spectra showing the extended absorption edge of TiO_2-$(MA)_2SnCl_4$ towards the visible region. (**b**) Tauc plots indicating band gap reduction in the nanocomposite. (**c**) The degradation of MB under sunlight, showing the superior photocatalytic efficiency of TiO_2-$(MA)_2SnCl_4$ film compared to TiO_2-NPs film and photolysis. (**d**) The proposed mechanism illustrating improved charge separation and electron transfer between TiO_2-NPs film and $(MA)_2SnCl_4$, facilitated by the heterojunction structure.

The presence of Sn^{2+} and Sn^{4+} induces OVs, which introduce additional electronic states within the TiO_2 band gap, effectively narrowing the band gap and improving visible-light absorption. The broader absorbance spectrum observed in the nanocomposite, compared to TiO_2-NPs, signifies improved light-harvesting capabilities, particularly in the visible range, which is advantageous for photocatalytic applications by enabling more efficient utilization of solar energy. The enhanced absorption is further supported by the formation of OVs during sunlight exposure [36,39], which generate localized energy levels, extending light absorption into the visible region and acting as active sites for catalytic reactions [39]. This combination of band gap narrowing, defect engineering, and extended absorption range significantly enhances the photocatalytic efficiency of the nanocomposite, making it a promising material for solar-driven applications.

The photocatalytic efficiency of TiO_2-NPs and $TiO_2/(MA)_2SnCl_4$ nanocomposite films was evaluated using methylene blue (MB) dye degradation under direct sunlight, assessing their ability to facilitate photodegradation. The TiO_2-NPs film exhibited moderate performance, achieving 75% MB degradation within 60 min, primarily due to its UV-light absorption, enabled by its band gap of 3.1 eV. In contrast, the $TiO_2/(MA)_2SnCl_4$ nanocomposite demonstrated superior photocatalytic activity, reaching 90% MB degradation within the same timeframe. Figure 5c highlights this enhanced efficiency, showing that the nanocomposite significantly outperforms TiO_2-NPs and the control experiment (photolysis without a catalyst), which exhibited negligible MB degradation, confirming the critical role of the photocatalyst in effective sunlight-driven degradation.

The improved photocatalytic performance of the $TiO_2/(MA)_2SnCl_4$ nanocomposite can be attributed to several key factors: (1) The incorporation of Sn and MA modifies the electronic structure, extending light absorption into the visible region and reducing the band gap to ~2.6 eV, enabling better solar energy utilization; (2) The formation of a Sn–O–Ti interaction within the nanocomposite facilitates charge transfer and efficient charge separation, minimizing electron–hole recombination and increasing the number of reactive charge carriers; and (3) The generation of OVs during sunlight exposure serves as active sites and electron traps, further enhancing photocatalytic activity by promoting interfacial charge transfer and surface redox reactions [45,46]. This synergistic combination of extended light absorption, effective charge separation, and active defect sites makes $TiO_2/(MA)_2SnCl_4$ a highly efficient material for solar-driven environmental remediation, offering significant advantages over conventional TiO_2-based photocatalysts.

The enhanced photocatalytic performance of the $TiO_2/(MA)_2SnCl_4$ nanocomposite, as illustrated in Figure 5d, results from a combination of synergistic effects, including extended light absorption, efficient charge separation, and structural modifications induced by sunlight exposure. The formation of Sn–O–Ti interactions within the nanocomposite facilitates strong interfacial bonding, significantly reducing the band gap to 2.6 eV compared to 3.1 eV for TiO_2-NPs. This reduction expands the absorption spectrum into the visible range, increasing the generation of photogenerated electron–hole pairs under solar irradiation. Upon sunlight exposure, both TiO_2 and Sn-containing species absorb photons, leading to electron excitation from the valence band to the conduction band, enabling efficient photocatalytic activity [20,21].

The band alignment between TiO_2 and Sn-containing phases promotes efficient charge transfer, where photoexcited electrons migrate to the conduction band of TiO_2, while holes move in the opposite direction, effectively minimizing recombination losses. This directional charge migration is driven by an internal electric field at the interface, a phenomenon observed in heterojunction-based photocatalysts such as S-scheme catalysts (e.g., N-doped CeO_{2}-δ@$ZnIn_2S_4$ and $TiO_2/g\text{-}C_3N_4$ heterojunctions) [47,48]. The internal charge redistribution ensures higher availability of charge carriers, which participate in surface redox reactions critical for pollutant degradation [45,46].

Furthermore, OVs generated during sunlight exposure play a pivotal role in enhancing charge carrier dynamics. These OVs serve as electron traps, further reducing electron–hole recombination, while also introducing intermediate states within the band gap, improving visible light absorption and adsorption of oxygen molecules and organic pollutants. During the photocatalytic reaction, photoexcited electrons in the TiO_2 conduction band reduce O_2 to superoxide radicals ($^{\bullet}O_2^-$), while holes in the valence band oxidize water or hydroxyl groups to generate hydroxyl radicals ($^{\bullet}OH$). These highly reactive species drive the degradation of pollutants such as MB, facilitating efficient photocatalytic performance.

The synergistic combination of band gap reduction, heterojunction-driven charge separation, OVs formation, and efficient reactive oxygen species (ROS) generation explains the superior photocatalytic activity of the $TiO_2/(MA)_2SnCl_4$ nanocomposite. These interactions enable the nanocomposite to outperform pure TiO_2 in organic pollutant degradation under sunlight, aligning with recent heterostructure-based charge transfer mechanisms that enhance photocatalytic efficiency.

3. Method Preparation and Characterization

3.1. Materials

The materials used in this study were utilized as received without any further purification. Titanium dioxide nanoparticles (TiO_2-NPs) were sourced as commercial TiO_2 P25 (99.5%, Evonik, Mount Waverley, VIC, Australia), known for exhibiting a small number of intrinsic defects. Methylammonium chloride (MA) (99%, Sigma-Aldrich, St. Louis, MO, USA), tin(II) chloride ($SnCl_2$) ($\geq$98%, Sigma-Aldrich, St. Louis, MO, USA), hydrochloric acid (HCl, 37%, Fisher Scientific, Hampton, NH, USA), and ethanol ($\geq$99.8%, Sigma-Aldrich, St. Louis, MO, USA) were also employed in the synthesis and preparation processes.

3.2. Preparation of MA and $SnCl_2$ Mixture

MA and $SnCl_2$ were dissolved in ethanol to prepare the precursor solution, maintaining a molar ratio of 2:1 for MA to $SnCl_2$. Before dissolving MA, HCl was added to the MA solution. This step was performed carefully to prevent over-acidification, ensuring that only a stoichiometric amount of HCl was used. After that, the $SnCl_2$ solution was added to the MA solution to form the MA and $SnCl_2$ mixture.

3.3. Synthesis of $TiO_2/(MA)_2SnCl_4$ Nanocomposite Film

TiO_2-NPs were dispersed in ethanol to form a stable colloidal suspension, which was then mixed with a 20% solution of MA and $SnCl_2$ precursor mixture at a 0.454:1 volume ratio. This mixture was drop-cast onto a pre-cleaned glass substrate and allowed to evaporate under ambient conditions, forming a uniform thin film. The resulting films were exposed to direct sunlight for 2 h to facilitate the reaction between MA and $SnCl_2$, resulting in the formation of $(MA)_2SnCl_4$ and the synthesis of the $TiO_2/(MA)_2SnCl_4$ nanocomposite film. This sunlight exposure was carried out under midday conditions in Al-Kharj City, Saudi Arabia. After sunlight exposure, the films were washed with ethanol to remove any unreacted precursors and dried under ambient conditions.

3.4. Characterization

The Raman spectroscopy was conducted using a SENTERRA II Compact Raman Microscope (Houston, TX, USA) with a 532 nm laser (2 mW power, 1 cm^{-1} resolution) to identify the vibrational modes of TiO_2-NPs and $TiO_2/(MA)_2SnCl_4$ films, providing insights into structural modifications and chemical interactions. Scanning electron microscopy (SEM) was performed with an FEI Quanta 250 instrument (Hillsboro, OR, USA) to examine surface morphology and nanoparticle distribution, revealing changes in film uniformity and particle agglomeration. Elemental composition and chemical states were analyzed through X-ray photoelectron spectroscopy (XPS) using a Thermo Scientific K-alpha system (Waltham, MA, USA) with an Al Kα source (1486.6 eV, 400 µm spot size), employing C 1s (284.5 eV) for charge correction to ensure accurate binding energy calibration. UV-Vis spectroscopy, conducted with a UV-5200 spectrophotometer (Shanghai Xiwen Biotech Co., Ltd., Shanghai, China), was used to assess the optical absorption properties of the films, evaluating their light-harvesting efficiency and band gap modifications.

3.5. Photocatalytic Performance Method

A 5 ppm aqueous solution of methylene blue (MB), used as a model organic pollutant, was prepared and placed in a stationary photocatalytic reactor and a magnetic stirrer set to 300 r.p.m. The $TiO_2/(MA)_2SnCl_4$ films, each with an area of 4 cm^2, were submerged 2 mm deep in 10 mL of the MB solution. To ensure adsorption/desorption equilibrium, the setup was kept in the dark for 30 min prior to sunlight exposure. The reactor was then exposed to direct sunlight, and control experiments using TiO_2-NPs films and no catalyst were conducted for comparison. Each experiment was repeated at least three times for reproducibility. Photocatalytic tests were carried out in midday sunlight in Al-Kharj City, Saudi Arabia. Samples were collected every 15 min during irradiation, and the concentration of MB was analyzed using UV-Vis spectroscopy by recording absorbance at 665 nm. The degradation efficiency was calculated using the following formula:

$$\text{Degradation efficiency (\%)} = (C_0 - C/C_0) \times 100$$

where C_0 is the initial concentration of the pollutant and C is the concentration at time t. This approach enabled a detailed comparison of the photocatalytic performance between the TiO_2-NPs and $TiO_2/(MA)_2SnCl_4$ films.

4. Conclusions

The $TiO_2/(MA)_2SnCl_4$ nanocomposite film developed in this study demonstrates significant advancements in photocatalytic performance, making it a promising material for water purification applications. The enhanced photocatalytic activity is attributed to the effective integration of Sn and MA components into the TiO_2-NPs framework, leading to extended light absorption into the visible region and improved charge separation via strong Sn–O–Ti interactions. UV-Vis spectroscopy confirmed a substantial band gap reduction from 3.1 eV in TiO_2-NPs to approximately 2.6 eV in the nanocomposite, improving visible-light utilization. XPS analysis further revealed the critical role of OVs, which facilitated charge trapping, minimized recombination, and enhanced catalytic efficiency. The $TiO_2/(MA)_2SnCl_4$ nanocomposite film achieved ~90% degradation of MB dye under sunlight within 60 min, outperforming the TiO_2-NPs film, which achieved only 75% degradation, underscoring the effectiveness of defect engineering and interfacial charge transfer. The synergistic effects of structural modification, optimized charge dynamics, and defect-mediated electronic interactions contribute to the nanocomposite's superior photocatalytic efficiency. The use of sunlight as a renewable energy source enhances sustainability, making this approach not only cost-effective and scalable but also a viable solution for addressing global water contamination challenges. These findings demonstrate that rational material design, incorporating controlled structural and electronic modifications, can significantly improve photocatalytic performance, paving the way for advanced environmental remediation technologies.

Author Contributions: Conceptualization, A.K. and T.F.Q.; Methodology, A.K. and A.M.A.; Software, A.O. and S.T.H.; Validation, N.A., S.T.H. and A.O.; Formal Analysis, A.M.A. and A.O.; Investigation, A.K., A.M.A. and N.A.; Resources, I.C.; Data Curation, I.C., S.T.H. and M.R.; experimental support, I.C.; experimental analysis, N.A.; Writing—Original Draft, A.K. and A.M.A.; Writing—Review and Editing, S.T.H., M.R. and T.F.Q.; Project Administration, T.F.Q.; Funding Acquisition, T.F.Q. All authors have read and agreed to the published version of the manuscript.

Funding: This project is sponsored by Prince Sattam Bin Abdulaziz University (PSAU) as a part of the funding for its SDG Roadmap Research Funding Programme project number PSAU-2023-SDG-94.

Data Availability Statement: Data are contained within the article.

Conflicts of Interest: The authors declare no conflict of interest.

References

1. Bolisetty, S.; Peydayesh, M.; Mezzenga, R. Sustainable Technologies for Water Purification from Heavy Metals: Review and Analysis. *Chem. Soc. Rev.* **2019**, *48*, 463–487. [CrossRef]
2. Ahmed, S.; Khan, F.S.A.; Mubarak, N.M.; Khalid, M.; Tan, Y.H.; Mazari, S.A.; Karri, R.R.; Abdullah, E.C. Emerging Pollutants and Their Removal Using Visible-Light Responsive Photocatalysis—A Comprehensive Review. *J. Environ. Chem. Eng.* **2021**, *9*, 106643. [CrossRef]
3. Mohamadpour, F.; Mohammad Amani, A. Photocatalytic Systems: Reactions, Mechanism, and Applications. *RSC Adv.* **2024**, *14*, 20609–20645. [CrossRef] [PubMed]
4. Mohapatra, L.; Yoo, S.H. Superwetting Materials as Catalysts in Photocatalysis: State-of-the-Art Review. *Chem. Eng. J.* **2024**, *481*, 148537. [CrossRef]
5. Chakravorty, A.; Roy, S. A Review of Photocatalysis, Basic Principles, Processes, and Materials. *Sustain. Chem. Environ.* **2024**, *8*, 100155. [CrossRef]
6. Armaković, S.J.; Savanović, M.M.; Armaković, S. Titanium Dioxide as the Most Used Photocatalyst for Water Purification: An Overview. *Catalysts* **2023**, *13*, 26. [CrossRef]
7. Dharma, H.N.C.; Jaafar, J.; Widiastuti, N.; Matsuyama, H.; Rajabsadeh, S.; Othman, M.H.D.; Rahman, M.A.; Jafri, N.N.M.; Suhaimin, N.S.; Nasir, A.M.; et al. A Review of Titanium Dioxide (TiO_2)-Based Photocatalyst for Oilfield-Produced Water Treatment. *Membranes* **2022**, *12*, 345. [CrossRef]
8. Lettieri, S.; Pavone, M.; Fioravanti, A.; Santamaria Amato, L.; Maddalena, P. Charge Carrier Processes and Optical Properties in TiO_2 and TiO_2-Based Heterojunction Photocatalysts: A Review. *Materials* **2021**, *14*, 1645. [CrossRef] [PubMed]
9. Magalhães, P.; Andrade, L.; Nunes, O.; Mendes, A. Titanium Dioxide Photocatalysis: Fundamentals and Application on Photoinactivation. *Rev. Adv. Mater. Sci.* **2017**, *51*, 91–129.
10. Du, S.; Lian, J.; Zhang, F. Visible Light-Responsive N-Doped TiO_2 Photocatalysis: Synthesis, Characterizations, and Applications. *Trans. Tianjin Univ.* **2022**, *28*, 33–52. [CrossRef]
11. Etacheri, V.; Di Valentin, C.; Schneider, J.; Bahnemann, D.; Pillai, S.C. Visible-Light Activation of TiO_2 Photocatalysts: Advances in Theory and Experiments. *J. Photochem. Photobiol. C Photochem. Rev.* **2015**, *25*, 1–29. [CrossRef]
12. Khan, H.; Shah, M.U.H. Modification Strategies of TiO_2 Based Photocatalysts for Enhanced Visible Light Activity and Energy Storage Ability: A Review. *J. Environ. Chem. Eng.* **2023**, *11*, 111532. [CrossRef]
13. Wajid Shah, M.; Zhu, Y.; Fan, X.; Zhao, J.; Li, Y.; Asim, S.; Wang, C. Facile Synthesis of Defective TiO_{2-x} Nanocrystals with High Surface Area and Tailoring Bandgap for Visible-Light Photocatalysis. *Sci. Rep.* **2015**, *5*, 15804. [CrossRef]
14. Wang, C.; Ding, Y.; Wang, Y.; Xie, Z.; Zeng, Z.; Li, X.; Ng, Y.H. Metal Halide Perovskites for Solar-to-Chemical Energy Conversion in Aqueous Media. *Carbon Energy* **2024**, *6*, e500. [CrossRef]
15. Purohit, S.; Yadav, K.L.; Satapathi, S. Metal Halide Perovskite Heterojunction for Photocatalytic Hydrogen Generation: Progress and Future Opportunities. *Adv. Mater. Interfaces* **2022**, *9*, 2200058. [CrossRef]
16. Mortan, C.; Hellmann, T.; Buchhorn, M.; Melzi d'Eril, M.; Clemens, O.; Mayer, T.; Jaegermann, W. Preparation of Methylammonium Lead Iodide ($CH_3NH_3PbI_3$) Thin Film Perovskite Solar Cells by Chemical Vapor Deposition Using Methylamine Gas (CH_3NH_2) and Hydrogen Iodide Gas. *Energy Sci. Eng.* **2024**, *8*, 3165–3173. [CrossRef]
17. Quarti, C.; Mosconi, E.; Ball, J.M.; D'Innocenzo, V.; Tao, C.; Pathak, S.; Snaith, H.J.; Petrozza, A.; Angelis, F.D. Structural and Optical Properties of Methylammonium Lead Iodide across the Tetragonal to Cubic Phase Transition: Implications for Perovskite Solar Cells. *Energy Environ. Sci.* **2016**, *9*, 155–163. [CrossRef]
18. Pan, L.; Feng, Y.; Kandlakunta, P.; Huang, J.; Cao, L.R. Performance of Perovskite CsPbBr3 Single Crystal Detector for Gamma-Ray Detection. *IEEE Trans. Nucl. Sci.* **2020**, *67*, 443–449. [CrossRef]

19. Baloch, A.; Hossain, M.; Tabet, N.; Alharbi, F. Practical Efficiency Limit of Methylammonium Lead Iodide Perovskite ($CH_3NH_3PbI_3$) Solar Cells. *J. Phys. Chem. Lett.* **2018**, *9*, 426–434. [CrossRef] [PubMed]
20. Cuzzupè, D.; Öz, S.; Ling, J.-K.; Illing, E.; Seewald, T.; Jose, R.; Olthof, S.; Fakharuddin, A.; Schmidt-Mende, L. Understanding Methylammonium Chloride-Assisted Crystallization for Improved Performance of Lead–Free Tin Perovskite Solar Cells. *Sol. RRL* **2023**, *7*, 2300770. [CrossRef]
21. Awais, M.; Kirsch, R.L.; Yeddu, V.; Saidaminov, M.I. Tin Halide Perovskites Going Forward: Frost Diagrams Offer Hints. *ACS Mater. Lett.* **2021**, *3*, 299–307. [CrossRef]
22. Kumar, A. Different Methods Used for the Synthesis of TiO_2 Based Nanomaterials: A Review. *Am. J. Nano Res. Appl.* **2018**, *6*, 1. [CrossRef]
23. Mani, S.S.; Rajendran, S.; Mathew, T.; Gopinath, S.C. A Review on the Recent Advances in the Design and Structure–Activity Relationship of TiO_2-Based Photocatalysts for Solar Hydrogen Production. *Energy Adv.* **2024**, *3*, 1472–1504. [CrossRef]
24. Khalid, A.; Ahmad, P.; Alharthi, A.; Muhammad, S.; Khandaker, M.; Faruque, M.R.; Faruque, I.; Din, I.U.; Alotaibi, M.; Kustov, L. Unmodified Titanium Dioxide Nanoparticles as a Potential Contrast Agent in Photon Emission Computed Tomography. *Crystals* **2021**, *11*, 171. [CrossRef]
25. Yadav, R.; Swain, D.; Kundu, P.P.; Nair, H.S.; Narayana, C.; Elizabeth, S. Dielectric and Raman Investigations of Structural Phase Transitions in $(C_2H_5NH_3)_2CdCl_4$. *Phys. Chem. Chem. Phys.* **2015**, *17*, 12207–12214. [CrossRef] [PubMed]
26. Tang, M.-C.; Zhang, S.; Magnanelli, T.; Nguyen, N.; Heilweil, E.; Anthopoulos, T.; Hacker, C. Unraveling the Compositional Heterogeneity and Carrier Dynamics of Alkali Cation Doped 3D/2D Perovskites with Improved Stability. *Mater. Adv.* **2020**, *2*, 1253–1262. [CrossRef]
27. Zhou, Q.; Liang, L.; Hu, J.; Cao, B.; Yang, L.; Wu, T.; Li, X.; Zhang, B.; Gao, P. High-Performance Perovskite Solar Cells with Enhanced Environmental Stability Based on a (p-FC6H4C2H4NH3)2[PbI4] Capping Layer. *Adv. Energy Mater.* **2019**, *9*, 1802595. [CrossRef]
28. Shao, Y.X.; Cai, Y.H.; Dong, D.; Wang, S.; Ang, S.G.; Xu, G.Q. Spectroscopic Study of Propargyl Chloride Attachment on Si(1 0 0) $- 2 \times 1$. *Chem. Phys. Lett.* **2009**, *482*, 77–80. [CrossRef]
29. Bandara, R.M.I.; Jayawardena, K.D.G.I.; Adeyemo, S.O.; Hinder, S.J.; Smith, J.A.; Thirimanne, H.M.; Wong, N.C.; Amin, F.M.; Freestone, B.G.; Parnell, A.J.; et al. Tin(IV) Dopant Removal through Anti-Solvent Engineering Enabling Tin Based Perovskite Solar Cells with High Charge Carrier Mobilities. *J. Mater. Chem. C* **2019**, *7*, 8389–8397. [CrossRef]
30. Alotaibi, M.A.; Alharthi, A.I.; Qahtan, T.F.; Alotibi, S.; Alansi, A.M.; Bakht, M.A. Photocatalytic Synthesis of Coumarin Derivatives Using Visible-Light-Responsive Strawberry Dye-Sensitized Titanium Dioxide Nanoparticles. *Catalysts* **2023**, *13*, 3001. [CrossRef]
31. Kang, S.; Mauchauffé, R.; You, Y.S.; Moon, S.Y. Insights into the Role of Plasma in Atmospheric Pressure Chemical Vapor Deposition of Titanium Dioxide Thin Films. *Sci. Rep.* **2018**, *8*, 16684. [CrossRef] [PubMed]
32. Chen, P.; Bai, Y.; Wang, S.; Lyu, M.; Yun, J.-H.; Wang, L. In Situ Growth of 2D Perovskite Capping Layer for Stable and Efficient Perovskite Solar Cells. *Adv. Funct. Mater.* **2018**, *28*, 1706923. [CrossRef]
33. Rocks, C.; Švrček, V.; Maguire, P.; Mariotti, D. Understanding Surface Chemistry during MAPbI 3 Spray Deposition and Its Effect on Photovoltaic Performance. *J. Mater. Chem. C* **2017**, *5*, 902–916. [CrossRef]
34. Xu, Y.; Wu, S.; Wan, P.; Sun, J.; Hood, Z. Introducing Ti^{3+} Defects Based on Lattice Distortion for Enhanced Visible Light Photoreactivity in TiO_2 Microspheres. *RSC Adv.* **2017**, *7*, 32461–32467. [CrossRef]
35. Li, Z.; Wang, S.; Wu, J.; Zhou, W. Recent Progress in Defective TiO_2 Photocatalysts for Energy and Environmental Applications. *Renew. Sustain. Energy Rev.* **2022**, *156*, 111980. [CrossRef]
36. Pan, X.; Yang, M.-Q.; Fu, X.; Zhang, N.; Xu, Y.-J. Defective TiO_2 with Oxygen Vacancies: Synthesis, Properties and Photocatalytic Applications. *Nanoscale* **2013**, *5*, 3601–3614. [CrossRef] [PubMed]
37. Yuan, Y.; Ji, Z.; Yan, G.; Li, Z.; Jinliang, L.; Kuang, M.; Jiang, B.; Zeng, L.; Pan, L.; Mai, W. TiO_2 Electron Transport Bilayer for All-Inorganic Perovskite Photodetectors with Remarkably Improved UV Stability toward Imaging Applications. *J. Mater. Sci. Technol.* **2021**, *75*, 39–47. [CrossRef]
38. Nguyen, D.; Ding, S.; Thuong, N.; Vân-Anh, N.; Mejía, E. Visible Light-Driven Degradation of Trichloroethylene in Aqueous Phase with Vanadium-Doped TiO 2 Photocatalysts. *Sol. RRL* **2023**, *7*. [CrossRef]
39. Guan, S.; Cheng, Y.; Hao, L.; Yoshida, H.; Tarashima, C.; Zhan, T.; Itoi, T.; Qiu, T.; Lu, Y. Oxygen Vacancies Induced Band Gap Narrowing for Efficient Visible-Light Response in Carbon-Doped TiO_2. *Sci. Rep.* **2023**, *13*, 14105. [CrossRef] [PubMed]
40. Chi, M.; Sun, X.; Lozano-Blanco, G.; Tatarchuk, B.J. XPS and FTIR Investigations of the Transient Photocatalytic Decomposition of Surface Carbon Contaminants from Anatase TiO_2 in UHV Starved Water/Oxygen Environments. *Appl. Surf. Sci.* **2021**, *570*, 151147. [CrossRef]
41. Alansi, A.M.; Qahtan, T.F.; Al Abass, N.; Al-Qunaibit, M.; Saleh, T.A. In-Situ Sunlight-Driven Tuning of Photo-Induced Electron-Hole Generation and Separation Rates in Bismuth Oxychlorobromide for Highly Efficient Water Decontamination under Visible Light Irradiation. *J. Colloid Interface Sci.* **2022**, *614*, 58–65. [CrossRef] [PubMed]

42. Bastug Azer, B.; Gulsaran, A.; Pennings, J.R.; Saritas, R.; Kocer, S.; Bennett, J.L.; Devdas Abhang, Y.; Pope, M.A.; Abdel-Rahman, E.; Yavuz, M. A Review: TiO_2 Based Photoelectrocatalytic Chemical Oxygen Demand Sensors and Their Usage in Industrial Applications. *J. Electroanal. Chem.* **2022**, *918*, 116466. [CrossRef]
43. Wang, Q.; Xiong, J.; Xing, Y.; Gan, X.; Zhu, W.; Xuan, R.; Liu, X.; Huang, L.; Zhu, Y.; Zhang, J. Reductive Sn Compensator for Efficient and Stable Sn-Pb Mixed Perovskite Solar Cells. *Adv. Sci.* **2024**, *11*, 2400962. [CrossRef] [PubMed]
44. Orudzhev, F.; Ramazanov, S.; Isaev, A.; Alikhanov, N.; Sobola, D.; Presniakov, M.Y.; Kaviyarasu, K. Self-Organization of Layered Perovskites on TiO_2 Nanotubes Surface by Atomic Layer Deposition. *Mater. Today Proc.* **2021**, *36*, 346–367. [CrossRef]
45. Yu, J.; Xiang, S.; Ge, M.; Zhang, Z.; Huang, J.; Tang, Y.; Sun, L.; Lin, C.; Lai, Y. Rational Construction of LaFeO3 Perovskite Nanoparticle-Modified TiO_2 Nanotube Arrays for Visible-Light Driven Photocatalytic Activity. *Coatings* **2018**, *8*, 374. [CrossRef]
46. Chen, C.; Zhou, J.; Geng, J.; Bao, R.; Wang, Z.; Xia, J.; Li, H. Perovskite $LaNiO_3/TiO_2$ Step-Scheme Heterojunction with Enhanced Photocatalytic Activity. *Appl. Surf. Sci.* **2020**, *503*, 144287. [CrossRef]
47. Farhan, S.; Hassan Raza, A.; Yang, S.; Yu, Z.; Wu, Y. Boosted Photocatalytic Hydrogen Evolution of S-Scheme N-Doped CeO2-δ@ZnIn2S4 Heterostructure Photocatalyst. *J. Colloid Interface Sci.* **2024**, *669*, 430–443. [CrossRef]
48. Yang, S.; Wang, K.; Chen, Q.; Wu, Y. Enhanced Photocatalytic Hydrogen Production of S-Scheme TiO_2/g-C3N4 Heterojunction Loaded with Single-Atom Ni. *J. Mater. Sci. Technol.* **2024**, *175*, 104–114. [CrossRef]

catalysts

MDPI

Article

Investigating the Photocatalytic Properties of Reduced Graphene Oxide-Coated Zirconium Dioxide and Their Impact on Structural and Morphological Features

Norhan Farghly [1], M. Abu El-Oyoun [1], A. Abousehly [1], Fatemah H. Alkallas [2], Amira Ben Gouider Trabelsi [2], E. R. Shaaban [3] and Abdelaziz Mohamed Aboraia [3,*]

[1] Physics Department, Faculty of Science, Assiut University, Assiut 7115, Egypt
[2] Department of Physics, College of Science, Princess Nourah Bint Abdulrahman University, P.O. Box 84428, Riyadh 11671, Saudi Arabia; aatrabelsi@pnu.edu.sa (A.B.G.T.)
[3] Department of Physics, Faculty of Science, Al-Azhar University, Assiut 71524, Egypt
* Correspondence: a.m.aboraia@gmail.com or a.m.aboraia@azhar.edu.eg

Abstract: Semiconductor photocatalytic technology demonstrates strong potential as a solution to defend environmental systems while converting energy. The photocatalytic behavior of traditional ZrO_2 catalysts suffers a major disadvantage because their activity remains low in visible light applications. XRD together with SEM, as well as EDX and EIS techniques, were utilized to evaluate the synthetic materials. This study demonstrated that the development of RGO-modified ZrO_2 heterostructures delivered substantial increases in photocatalytic functionality through effective photogenerated charge separation mechanisms. Tests showed the RGO/ZrO_2 heterostructures exhibited outstanding photocatalytic behavior that led to an 80% MB solution breakdown in 120 min while exceeding electrocatalytic parameters in multiple tests. The experimental data from UV–vis spectroscopy combined with electrochemical analysis and radical trapping methods demonstrated that heterostructure improvement resulted from higher light absorption rates and effective active site exposure while providing better electron/hole pair separation. This research establishes S-scheme heterostructures to enable advancements in environmental protection alongside energy conversion technologies.

Keywords: ZrO_2 nanoparticles; reduced graphene oxide; photocatalysis; Mott–Schottky; MB

1. Introduction

Recently, ecologists have focused their attention on water-soluble and volatile compounds, especially dyes [1,2]. Wastes from the dye and textile industries are very harmful to aquatic life [3,4]. These residues represent a substantial danger to human health and can cause cancer. The release of dyes into bodies of water is the primary cause of eutrophication, the contamination of water, and aquatic ecosystem disruption [5,6]. Extensive efforts in the last few years to tackle these environmental issues have focused mostly on the photocatalytic process as the best method for successful dye repair in the presence of light [7,8]. The study of photocatalytic techniques is closely linked to the examination of materials used in photocatalysis. ZrO_2 is a transition metal oxide that is environmentally friendly, as it has low toxicity and is chemically stable [9,10]. Furthermore, it is biobased, inexpensive, and exhibits outstanding thermal stability as well as electrochemical characteristics [11,12]. Because of its exceptional electrocatalytic characteristics, It works well as an active electrode material in crucial electrocatalytic investigations [13,14]. Currently, semiconductor

https://doi.org/10.3390/catal15030289

elements, such as ZnO, TiO_2, and ZrO_2, are now the most common materials utilized in photocatalysts, serving as the catalyst's primary component [15–17]. Among these, ZrO_2 is known as the only metal having oxidizing and reducing properties due to its acidic nature and properties [18].

Since ZrO_2 is sorted as a semiconductor with the p-type, it readily generates oxygen holes, which can carry excitation and promote the interaction between active layers [19–21]. Consequently, it shows improved viability as a photocatalyst when contrasted with elective materials [9,22]. The zirconia experiences difficulties when utilized as an independent semiconductor, for example, a high pace of transporter recombination and diminished quantum proficiency, which to some degree limits its down-to-earth utility [23,24]. The broadband hole limitation of zirconia materials restricts their ability to absorb UV radiation while blocking out nearly 96% of sunlight across the spectrum. Thus, its far-reaching application is restricted [23,25]. Currently, numerous research studies center on improving the photocatalytic effectiveness of zirconia through various techniques, including doping, incorporating carbon nanotubes, and employing several other methods [26,27]. Researchers have found the last method to be the most effective, resulting in the creation of composites by various scientists in order to enhance their photocatalytic activity in visible light [28–30]. Previous studies have shown that when exposed to UV light, ZrO_2 can effectively break down textile dyes, for instance, rhodamine B, reactive yellow, and 2,2,4-diphenoxyacetic through a process known as photodegradation [31]. The improved reactivity observed in the UV light in the previous examples can be attributed to two factors, either the catalyst's stability or an enlargement of the surface area. Another contributing factor is the charge transfer process facilitated by ZrO_2, which benefits from its small particle size, large surface area, and a large number of surface hydroxyl (OH) groups. In recent times, there has been significant emphasis from researchers on enhancing the photocatalytic capabilities of carbon-based nanocomposites [32]. As a result, graphene, an exceptionally appealing component, was specially chosen for the production of these composites [33]. This is attributed to its large specific surface area and a two-dimensional p-p conjugation structure with outstanding electron conductivity [34]. Previous research has examined the characteristics of transporters in pure graphene layers, revealing that they exhibit similarities to massless Dirac fermions. As an example, when TiO_2 nanocrystals are mixed with graphene, electrons from the band of conduction (CB) that are stimulated by TiO_2 can transfer to go by means of a percolation procedure [35]. The Schottky barrier is a heterojunction generated at the interface that effectively distinguishes photo-induced electron–hole pairs, limiting charge recombination [36]. Several techniques have been used to produce RGO; in our investigation, the RGO is produced using waste bottle plastics. ZrO_2 was covered with RGO using an affordable and efficient process. The morphological and structural qualities were examined. For the first time, electrocatalysis and photocatalysis were investigated in ZrO_2@%RGO.

2. Result and Discussion

Figure 1 displays the XRD patterns of pure ZrO_2 and ZrO_2 covered with various RGO concentrations. It was found by X'pert HighScore search–match analysis that the tetragonal phase of ZrO_2 "(P 42/n m c:1) is present. The widening of peaks observed in all the patterns indicates that the powders synthesized have a noncrystalline structure. Coated samples indicate that RGO has an amorphous phase compared to the pure sample. Moreover, the distinct and sharp diffraction peaks clearly support the successful synthesis to pure ZrO_2 tetragonal through calcination. Furthermore, for the remaining groups, significantly different peaks were recorded at "28.2°, 31.5°, 34.1°, 35.3°, and 50.1°, which can be directly attributable to the tetragonal ZrO_2".

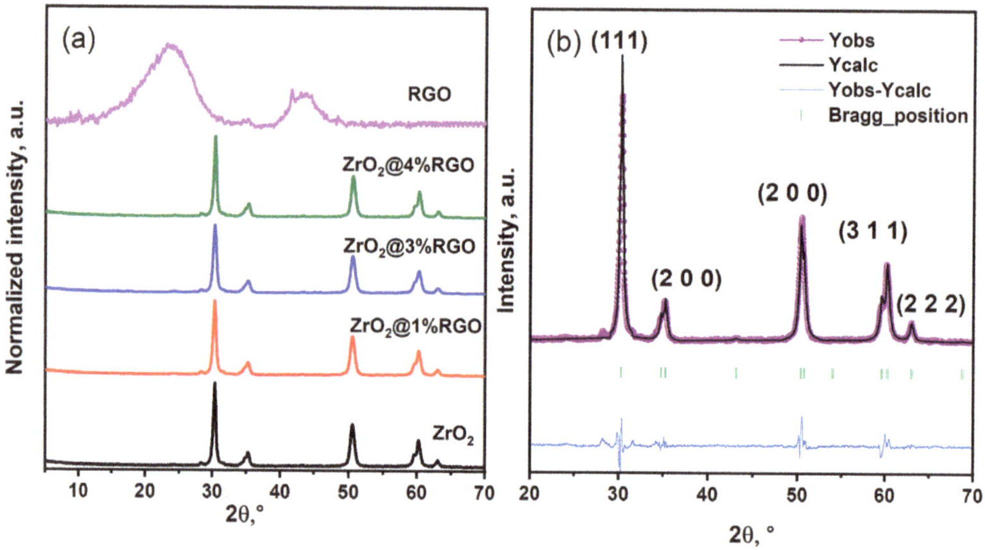

Figure 1. (**a**) The XRD analysis displayed patterns for ZrO_2, RGO and ZrO_2 covered with various amounts of RGO. (**b**) The rietveld refinement of pure ZrO_2.

SEM images observed from pure ZrO_2 and covered with varying percentages of RGO nanocomposites are shown in Figure 2. The pure ZrO_2 demonstrates a surface that is smooth and does not have any additional substances added to it, as demonstrated in Figure 2a. However, Figure 2b,c shows SEM images of the RGO-ZrO_2 nanocomposite, which demonstrates the formation of a several-layer structure made up of RGO sheets. Furthermore, larger accumulated ZrO_2 particles were seen on the surface of the RGO sheets. The created ZrO_2 has an irregular shape, with particle sizes less than 100 nm. This combination results from heterogeneous solid nucleation among RGO and ZrO_2. Figure 3a shows the results of EDAX analysis, which was used to determine the elements present in the nanocomposite. EDAX data analysis "showed Zr, C, and O contents of 64, 25.2, and 10.0%", correspondingly. The elements C, O, and Zr, in this distribution, grant confirmation for the formation of RGO-ZrO_2 nanocomposites. When a single layer of RGO is combined with a metal oxide to form a nanocomposite, a multilayer structure is formed. Additionally, due to the limitations of SEM, it can also be viewed as multidimensional.

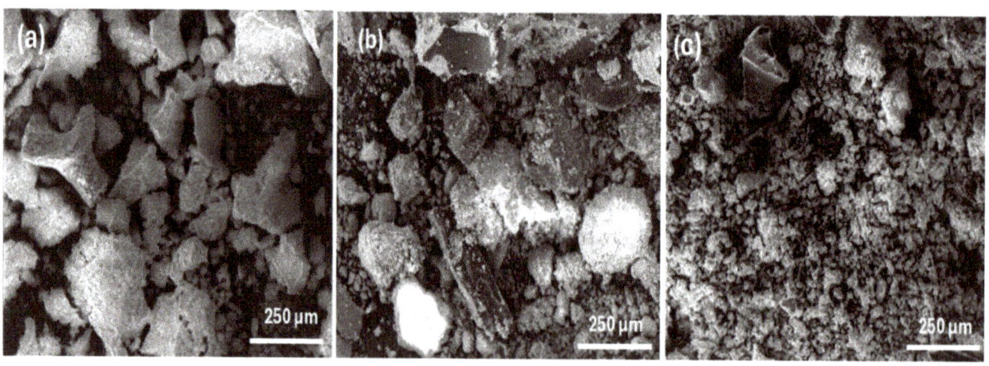

Figure 2. (**a**) SEM photo of pure ZrO_2, (**b**) ZrO_2@1%rGO, and (**c**) ZrO_2@4%rGO.

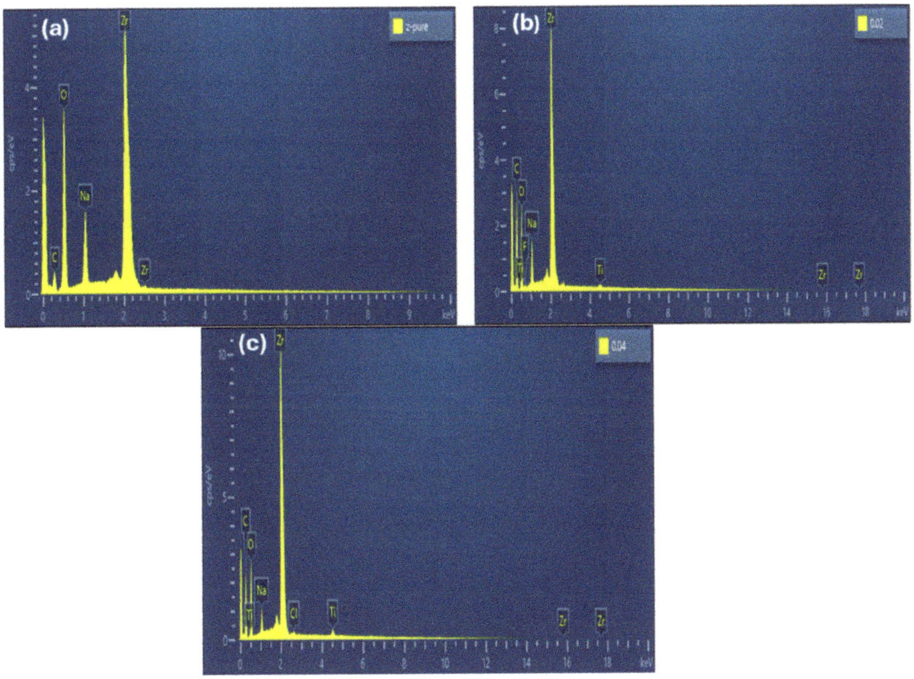

Figure 3. (**a**) EDX of pure ZrO_2, (**b**) ZrO_2@1%rGO, and (**c**) ZrO_2@4%rGO.

The band gap "Eg" that characterizes the electronic configuration of both pristine ZrO_2 and covered ZrO_2 with various amounts of RGO is determined by the lowest possible energy transfer between their valence and conduction bands. The metal oxide nanoparticles triggered the production of reactive oxygen species (ROS) in the photocatalysis mechanism. The creation of ROS—such as $\cdot OH^-$ and O_2 that occur using ZrO_2-coated ZrO_2 with different concentrations of RGO materials—depends on the energy band gap and redox potential of the variations among reactive species. The energy band gap (E_g) value is extracted from the extrapolation of the linear section of a plot of $(\alpha.h\nu)^2$ vs. photon energy (eV), as presented in Figure 4. "The energy band gap (Eg) values for ZrO_2 nanoparticles and ZrO_2 coated with different concentrations of RGO nanocomposites were decreased from 5.6 and 3.9 eV, respectively. The energy band gap value obtained for ZrO_2/RGO closely matched the value reported by Braj Raj Singh and colleagues [37]." Compared to ZrO_2, ZrO_2/RGO has a lower energy band and displays a higher electronic surface area, indicating its potential as a photoconductor that is more active when exposed to sunlight. The low surface area of ZrO_2/RGO is due to its good crystallinity and lack of oxygen vacancies. One possible explanation for the band gap lowering is the chemical bonding among ZrO_2 and certain carbon sites through the production method. When ZrO_2 was placed on the prepared RGO sheets, the Eg value decreased to 3.86 eV, showing that ZrO_2@%RGO can absorb visible light. The decrease in band ratio can be ascribed to the Zr-C and Zr-O-C bonds among ZrO_2 and RGO. These results indicate that ZrO_2@%RGO nanocomposites "NCs" have good photocatalytic activity.

The flat band potential of each semiconductor was determined using Mott–Schottky analysis as described in Supplementary Information. The electrical band structures of pure ZrO_2 and ZrO_2@%RGO were investigated and shown in Figure 5a–d. Based on the plot features, it is possible to conclude that the positive slope found in both pure ZrO_2 and ZrO_2@%RGO indicates the presence of n-type behavior. The Mott–Schottky performance of all semiconductor devices was analyzed at a frequency of 500 Hz. "As-synthesized photocatalysts' flat band potential is 0.17, 0.20, 0.53, and 0.29 eV vs. SCE for ZrO_2 and coated with different concentrations of 0.01, 0.03, and 0.04 RGO, correspondingly."

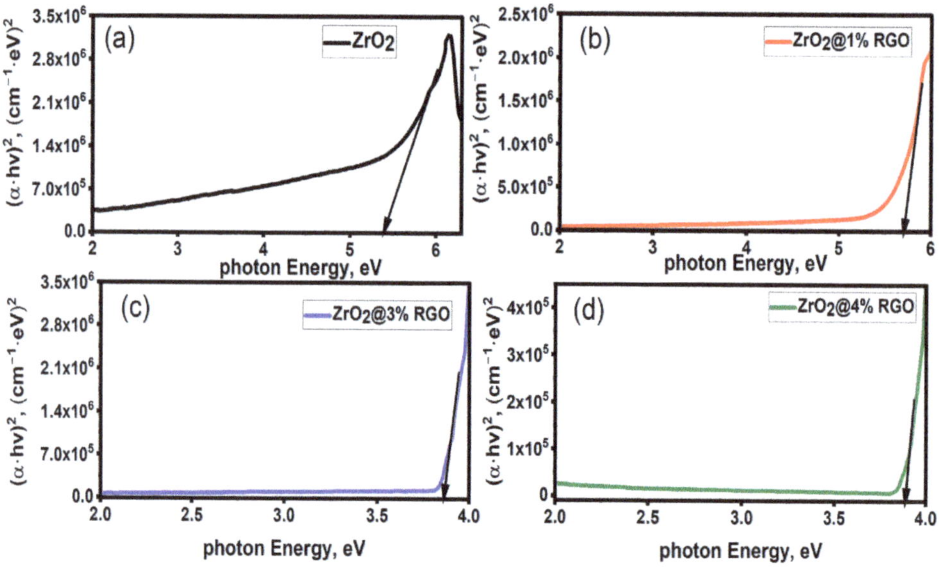

Figure 4. The optical energy gap of (**a**) ZrO_2, (**b**) ZrO_2@1%RGO, (**c**) ZrO_2@3%RGO, and (**d**) ZrO_2@4%RGO.

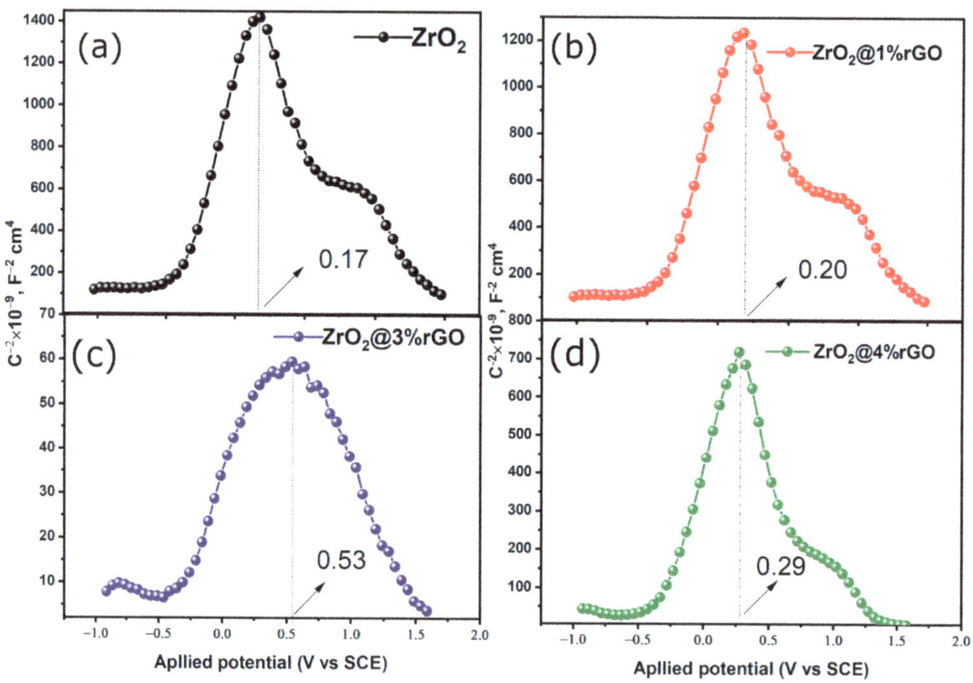

Figure 5. Mott–Schottky plots of (**a**) ZrO_2, (**b**) ZrO_2@1%RGO, (**c**) ZrO_2@3%RGO, and (**d**) ZrO_2@4%RGO.

The effectiveness of photogenerated exciton separation and the resistance of the interface layer in pristine ZrO_2 and ZrO_2@%RGO photocatalysts was evaluated utilizing EIS, as depicted in Figure 6, depicting the electrolyte system due to the electrolyte and electronic resistance of the photoelectrodes. The low-frequency portion pertains to the Warburg resistance. It is noted that ZrO_2@4% RGO is smaller compared to pure ZrO_2 and other photocatalysts, suggesting its effective interfacial charge separation and superior conductivity characteristics. The high electrical conductivity facilitates electron flow at the electrode–electrolyte solution interface. The preceding analysis confirms the crucial contribution of ZrO_2 in the composite to prolong the lifespan of photoinduced charge carriers.

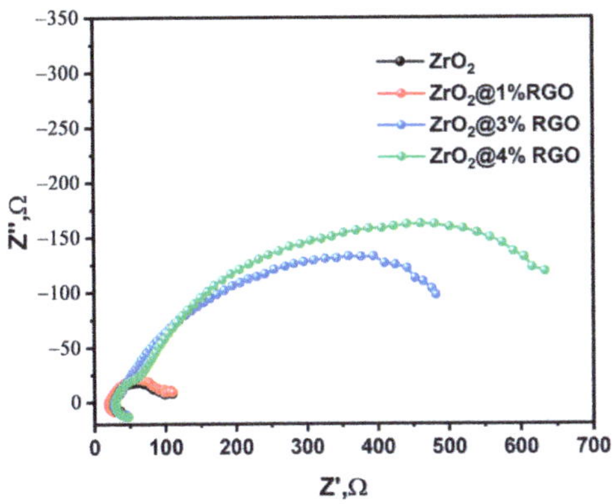

Figure 6. Electrochemical impedance (EIS) spectra of pristine ZrO_2, ZrO_2@1%RGO, ZrO_2@3%RGO, and ZrO_2@4%RGO.

Photocatalytic Performance

The study of the optical characteristics of a photocatalyst is largely defined by the ability to promote the absorption of UV-Vis light. There is no significant change in the dye degradation without a catalyst as, demonstrated in Figure S2. UV-Vis absorption spectra of ZrO_2 nanoparticles (NM) and ZrO_2@%RGO composites are given in Figure 7a–d. The photocatalytic activity of synthesized ZrO_2 and ZrO_2@%RGO was analyzed by checking the decolorization of methylene blue (MB) dye solution exposed to 120 min of sunlight. The results showed that the ZrO_2@%RGO sample had higher photocatalytic activity than ZrO_2 because of the better sunlight absorption and fewer electron–hole recombination rates in the metal oxides due to the incorporation of reduced graphene oxide. The photocatalytic tests were conducted after the dye molecules were absorbed into the composite material to the greatest extent. When this concentration was associated with light, minimal degradation for the dye was recorded when the solution was in the dark. Likewise, the percentage of MB changed very little when the sample was put under sunlight in the absence of the photocatalyst. In the final 120 min after the samples had been exposed to sunlight, there was negligible photolysis of the dye. The degradation efficiencies of MB for ZrO_2 and ZrO_2@%RGO are demonstrated at Figure 8a under sunlight irradiation of 65% and 80%, respectively. In Figure 8b, the photocatalytic decomposition of MB between ZrO_2 and ZrO_2@%RGO under sunlight irradiation was compared. "Furthermore, the degradation kinetics of MB have been modelled and plotted based on the principles of pseudo-first-order kinetics theory ($\ln (C_0/C_t)) = kt$, where k represents the apparent first-order rate constant, as shown in Figure 8b." It may be shown that the rate constants for ZrO_2, ZrO_2@1%RGO, ZrO_2@3%RGO, and ZrO_2@4%RGO are 0.00886, 0.00975, 0.00999, and 0.0119 min^{-1}, respectively. The ZrO_2@4%RGO sample exhibits the greatest k value, which is around twice as large as that of ZrO_2.

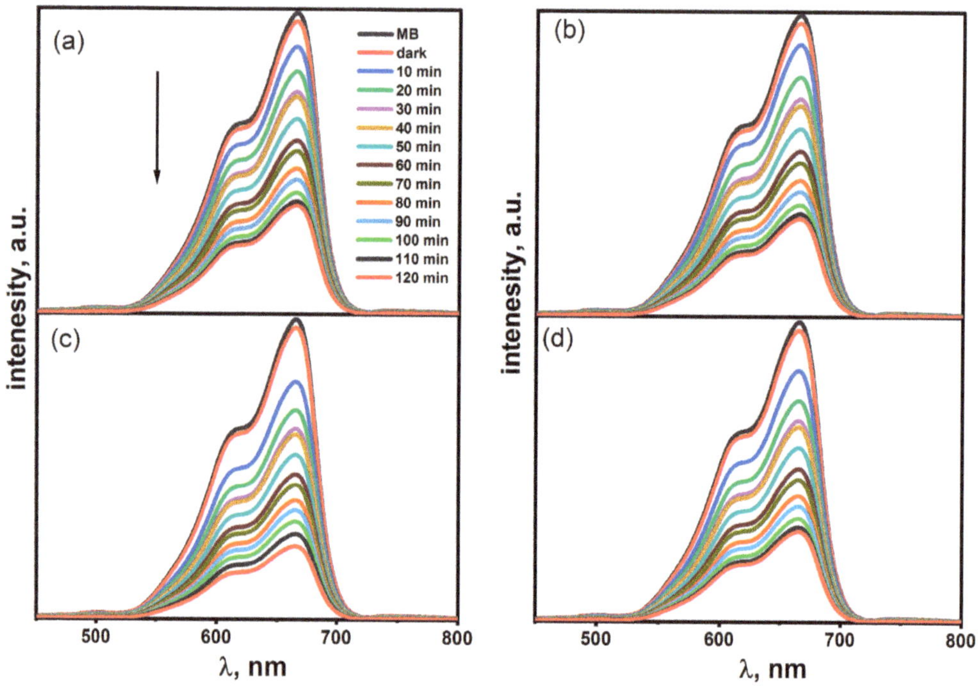

Figure 7. Photocatalytic degradation performance for (**a**) pristine ZrO_2, (**b**) ZrO_2@1%RGO, (**c**) ZrO_2@3%RGO, and (**d**) ZrO_2@4%RGO.

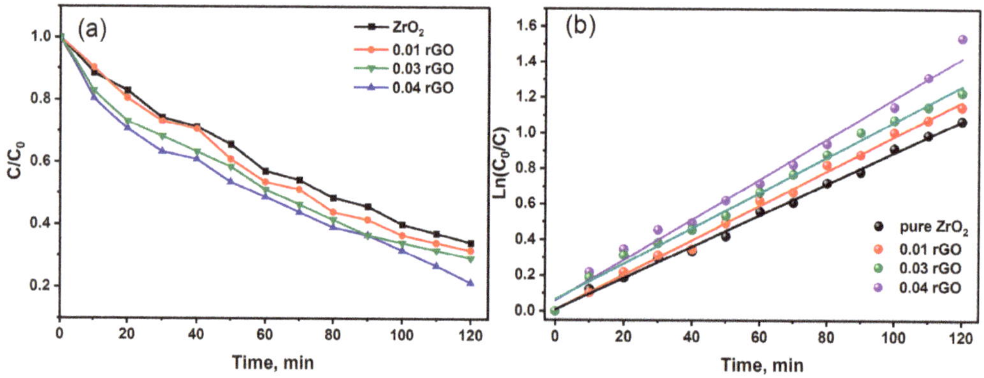

Figure 8. (**a**) Photocatalytic efficiency of pure ZrO_2, ZrO_2@1%RGO, ZrO_2@3%RGO, and ZrO_2@4%RGO and (**b**) $Ln(C_o/C)$ vs. time for pure ZrO_2, ZrO_2@1%RGO, ZrO_2@3%RGO, and ZrO_2@4%RGO.

Light exposure failed to initiate significant dye breakdown in the observations. Methylene blue (MB) showed a minimal photolysis reaction in the absence of a photocatalyst under sunlight exposure which lasted up to 120 min. Figure 9a displays MB degradation levels at 65% with ZrO_2 and 80% with ZrO_2/rGO composite under sunlight irradiation. The light stability of these photocatalysts was assessed through recurring tests under sunlight irradiation, as shown in Figure 9b. For ZrO_2@4%rGO, the recycling test was performed over four cycles using the same sample with a fresh dose of dye introduced for degradation in each cycle. Between cycles, the samples were washed and dried prior to the next run.

Over the first three cycles, the degradation percentage remained consistent; however, after the fourth cycle, there was a decrease in degradation efficiency, likely due to sample loss during washing. This confirms that ZrO_2@4%rGO exhibits excellent recycling efficiency and maintains stability throughout the process. Table 1 shows a comparative evaluation of ZrO_2 structures that degrade MB dye through photocatalysis.

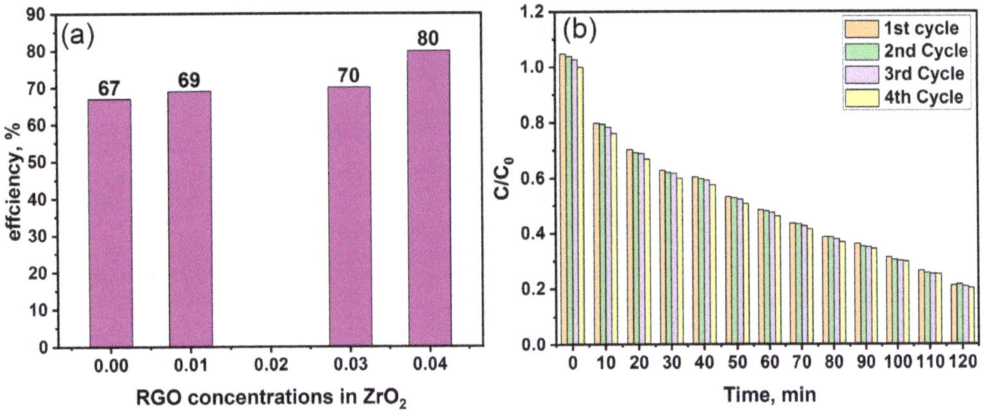

Figure 9. (**a**) Degradation rate of pristine ZrO_2, ZrO_2@1%RGO, ZrO_2@3%RGO, and ZrO_2@4%RGO. (**b**) Stability of the ZrO_2@4%RGO for MB removal.

Table 1. A comparative evaluation of ZrO_2 structures that degrade MB dye through photocatalysts.

Catalyst	Synthesis Method	Dye	Irradiation Source	Efficiency, %	References
t-ZrO_2	Green method	MB	UV light	68	[38]
ZrO_2	Biogenic	MO	UV light	69	[39]
Nd-ZrO_2	Polymer assisted	MB	Visible light	68	[40]
ZrO_2 NPs	Hydrothermal	MB	Sunlight	80	[41]
ZrO_2	Sol–gel combustion	MB	Sunlight	65	Our work
ZrO_2@4%RGO	Sol–gel combustion	MB	Sunlight	82	Our work

The process of photodegradation for MB dye on the surface of a photocatalyst entails the creation of electron–hole pairs. To enhance the photocatalytic activity of the ZrO_2/RGO photocatalyst, it needs to absorb photons from UV-Vis light irradiation. The energy absorbed by the photocatalyst should be equal to or greater than its band gap energy. Consequently, this results in the excitation of electrons moving from the valence band (VB) to the conduction band (CB), leaving behind holes in the VB and generating electrons in the CB. The conduction band (CB) edge level of ZrO_2 is higher than the Fermi energy of RGO, causing excited photoelectrons to transfer from the CB of ZrO_2 to the surface of RGO. This movement of electrons helps RGO reduce the recombination rate of photoinduced charge carriers and prolongs the lifetime of the electron–hole pairs. As a result, RGO significantly enhances the photocatalytic activity of ZrO_2. The electrons transferred to rGO can then be captured by dissolved O_2 molecules, forming superoxide radicals ($O_2^{-}\bullet$). Meanwhile, the photo-induced holes in ZrO_2 may react with adjacent hydroxyl groups (OH^-) or adsorbed H_2O molecules to produce hydroxyl radicals ($\bullet OH$). These reactive oxygen species are responsible for the photodegradation of MB, as illustrated in Figure 10.

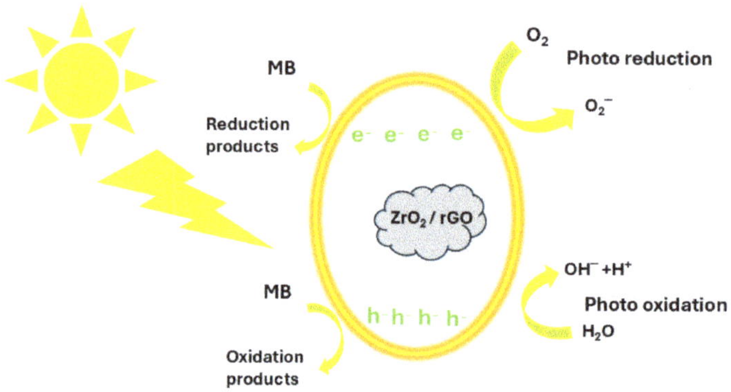

Figure 10. Reported process for ZrO_2/RGO photocatalyzed degradation of MB.

Figure 11 depicts the extent of the photocatalytic degradation of MB dye both with and without scavengers. In the absence of any quencher, the dye degradation achieved 84%. However, this percentage reduced to 58% and 48% when P-BQ and AO were present, respectively. Notably, the degradation of MB significantly declined to 19% in the presence of t-BuoH, suggesting that electrons play a main role in the photodegradation of MB dye through the formation of $^{\cdot}O_2{}^{-}$ radicals.

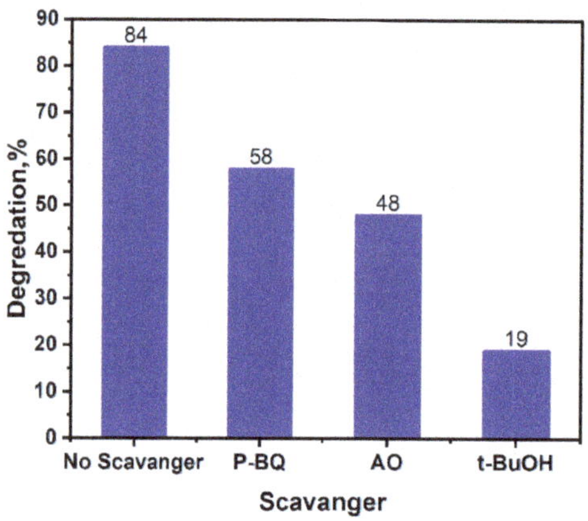

Figure 11. The influence of different scavengers on the photocatalytic degradation of MB.

3. Experimental Work

3.1. Preparation Technique of ZrO_2

Zirconium dioxide (ZrO_2) nanoparticles were made using a method called the combustion of sol–gel. All the chemicals used in this study, including citric acetate, zirconium nitrate, and DMF, were obtained from Sigma Aldrich, St. Saint Louis, MO, USA. To maintain a consistent citric acid-to-metal cation ratio, the solution was thoroughly mixed at room temperature. To remove the solvent and form a gel, the solution was vigorously stirred at 150 °C while being frequently swirled. The gel precipitate was centrifuged and repeatedly cleaned with ethanol and water in order to cleanse it. Finally, it was heated to 650 °C and ground into a fine powder. Compared to other techniques, sol–gel combustion is a more attractive choice for material synthesis because it is easy to implement, requires minimal resources, and delivers high yields. Because it minimizes waste and maximizes resource utilization, the sol–gel combustion technique complements the closed-loop approach of the circular financial system.

3.1.1. Reduced Graphene Preparation from PET Water Bottle Waste

Wastewater bottles composed of polyethylene terephthalate (PET) were mechanically broken down and sieved using a standard sieve shaker to obtain particles within the size range of 1–3 mm. A 2 g portion of the untreated PET waste was then placed into a 50 mL stainless steel autoclave reactor (SS316). The sealed reactor was positioned in the center of an oven and heated at a rate of 8 °C/min until it reached 800 °C, a temperature maintained for 1 h. Following the annealing treatment, the system was permitted to undergo spontaneous cooling over a 12-h periods. The resulting dark product was collected, characterized, and subsequently crushed. The yield is RGO as shown in Figure S1. The PET waste bottles acted as a precursor material for graphene fabrication, as depicted in Scheme 1.

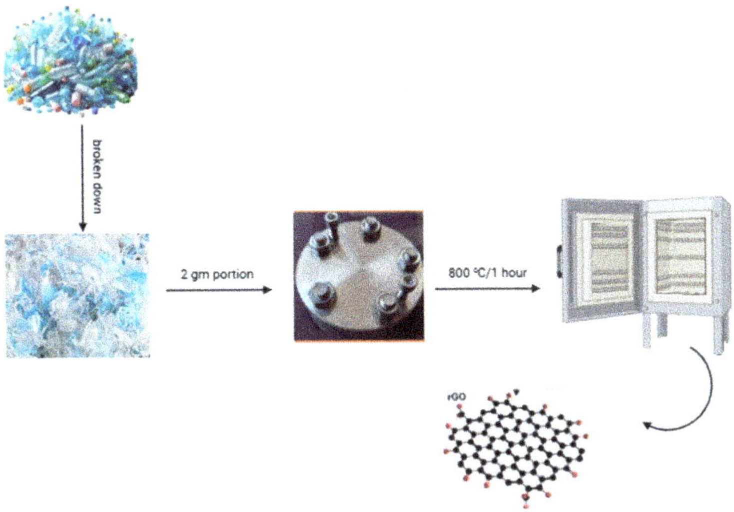

Scheme 1. The PET waste bottles acted as a precursor material for reduced graphene fabrication.

3.1.2. Synthesis ZrO_2@%RGO

Photocatalysts ZrO_2@%RGO were produced through an ultrasonic mixing technique. 0.05 g of ZrO_2 was disappeared in 30 DI water for several minutes. Following the addition of different concentrations of RGO (0, 0.01, 0.03, and 0.04) and 1 h sonication, the ZrO_2 mixture was centrifuged, cleaned, and dried (60 °C, 24 h) to yield ZrO_2@%RGO.

3.2. Characterization

Produced via a straightforward chemical route (sol–gel combustion), the nanoparticles were probed employing "a Philips PW 1710 X-ray" diffractometer with "CuKα radiation. (λ = 1.5405 Å)", EDAX, and SEM). The JASCO 670 UV-Vis double-beam spectrophotometer was used to measure the transmittance spectra. Characterization was performed across the UV-Vis spectrum, from 200 nm to 1000 nm. The degradation of MB was used to test the materials' photocatalytic capability, a representative organic pollutant. Degradation of the catalyst under light irradiation (photodegradability) was assessed using UV–visible spectroscopy. An analysis of impedance and Mott–Schottky measurements occurred through the Cortest Cs305 workstation system. A 500-watt xenon lamp worked together with 25 mg of photocatalyst material for conducting the photocatalysis experiment.

4. Conclusions

In summary, the ZrO_2 coated with various concentrations of RGO compounds is successfully made by the sonication technique. The XRD analysis via X'pert HighScore search reveals that the tetragonal phase of ZrO_2 (P 42/nm c:1) is present in all the concentrations of RGO. The peak broadening of all patterns is visible for the synthesized powders when compared to the pure ones, which indicates that they are all noncrystalline. The coated sample relates to the amorphous phase of RGO. The pure ZrO_2 without any additives exhibits a smooth surface when compared to the coating of RGO. The multilayer structure of rGO sheets is clearly seen in the SEM image of the RGO-ZrO_2 nanocomposite. A sequence of larger aggregated ZrO_2 particles is seen on the surface of RGO. The catalytic efficiency of ZrO_2 is improved by the formation of ZrO_2/RGO composites. Studies revealed ZrO_2@4%RGO demonstrated the maximum photocatalytic efficiency among all the weight percentage tests for ZrO_2/RGO. The nanocomposite presents a reduced bandgap value that falls from 5.6 eV to 3.87 eV. The Mott–Schottky performance of all semiconductor devices was analyzed at a frequency of 500 Hz. As-synthesized photocatalysts' flat band potential is 17×10^{-2}, 20×10^{-2}, 53×10^{-2}, and 29×10^{-2} eV vs. SCE for pristine ZrO_2 and covered with different amount of 0.01, 0.03, and 0.04 RGO, correspondingly. The ZrO_2@4%RGO composite shows a 1.5 times higher MB photocatalytic rate because its reduced band gap allows for effective charge carrier separation and enables enhanced charge transfer through increased surface area. The reduced graphene oxide nanocomposites prove to be ideal materials for sustainable water pollution remediation. The debasement efficiencies of MB utilizing ZrO_2 and ZrO_2@%RGO under sunshine irradiation were estimated at 65% and 80%, respectively.

Supplementary Materials: The following supporting information can be downloaded at: https://www.mdpi.com/article/10.3390/catal15030289/s1, Figure S1: The XRD patterns of RGO and SEM photo of RGO (in set), Figure S2: The Uv-vis spectra of MB in 120 min.

Author Contributions: Conceptualization, N.F. and M.A.E.-O.; methodology, A.A.; software, F.H.A.; validation, A.B.G.T.; formal analysis, E.R.S.; investigation, resources, and data curation, A.M.A.; writing—original draft preparation, all authors. All authors have read and agreed to the published version of the manuscript.

Funding: This research was funded by Princess Nourah bint Abdulrahman University Researchers Supporting Project number (PNURSP2025R38), Princess Nourah bint Abdulrahman University, Riyadh, Saudi Arabia.

Data Availability Statement: The data will be available upon reasonable request.

Acknowledgments: The authors express their gratitude to Princess Nourah bint Abdulrahman University Researchers Supporting Project number (PNURSP2025R38), Princess Nourah bint Abdulrahman University, Riyadh, Saudi Arabia.

Conflicts of Interest: The authors declare no conflicts of interest.

References

1. Liu, T.; Aniagor, C.O.; Ejimofor, M.I.; Menkiti, M.C.; Wakawa, Y.M.; Li, J.; Akbour, R.A.; Yap, P.-S.; Lau, S.Y.; Jeevanandam, J. Recent developments in the utilization of modified graphene oxide to adsorb dyes from water: A review. *J. Ind. Eng. Chem.* **2023**, *117*, 21–37. [CrossRef]
2. Aboraia, A.M.; Almohammedi, A.; Alraddadi, S.; Taha, S.A.; Saad, M.; Sharaf, I.; Ismail, Y.A.M. Comparison study between as-synthesized ZnO and ZnO derived from ZiF-8 metalorganic framework in removing methylene blue. *Mod. Phys. Lett. B* **2024**, *38*, 2450221. [CrossRef]
3. Al-Tohamy, R.; Ali, S.S.; Li, F.; Okasha, K.M.; Mahmoud, Y.A.G.; Elsamahy, T.; Jiao, H.; Fu, Y.; Sun, J. A critical review on the treatment of dye-containing wastewater: Ecotoxicological and health concerns of textile dyes and possible remediation approaches for environmental safety. *Ecotoxicol. Environ. Saf.* **2022**, *231*, 113160. [CrossRef]

4. Khattab, T.A.; Abdelrahman, M.S.; Rehan, M. Textile dyeing industry: Environmental impacts and remediation. *Environ. Sci. Pollut. Res.* **2020**, *27*, 3803–3818. [CrossRef] [PubMed]
5. Sharma, J.; Sharma, S.; Soni, V. Classification and impact of synthetic textile dyes on Aquatic Flora: A review. *Reg. Stud. Mar. Sci.* **2021**, *45*, 101802. [CrossRef]
6. Elgarahy, A.M.; Elwakeel, K.Z.; Mohammad, S.H.; Elshoubaky, G.A. A critical review of biosorption of dyes, heavy metals and metalloids from wastewater as an efficient and green process. *Clean. Eng. Technol.* **2021**, *4*, 100209. [CrossRef]
7. Ibrahim, I.A.; Rawindran, H.; Alam, M.M.; Leong, W.H.; Sahrin, N.T.; Ng, H.-S.; Chan, Y.J.; Abdelfattah, E.A.; Lim, J.W.; Aliyu, U.S.; et al. Mitigating persistent organic pollutants from marine plastics through enhanced recycling: A review. *Environ. Res.* **2023**, *240*, 117533. [CrossRef]
8. Aboraia, A.M.; Al-omoush, M.; Solayman, M.; Saad, H.M.H.; Khabiri, G.; Saad, M.; Alsulaim, G.M.; Soldatov, A.V.; Ismail, Y.A.M.; Gomaa, H. A heterostructural MoS_2QDs@UiO-66 nanocomposite for the highly efficient photocatalytic degradation of methylene blue under visible light and simulated sunlight. *RSC Adv.* **2023**, *13*, 34598–34609. [CrossRef]
9. Sakfali, J.; Chaabene, S.B.; Akkari, R.; Dappozze, F.; Berhault, G.; Guillard, C.; Zina, M.S. High photocatalytic activity of aerogel tetragonal and monoclinic ZrO_2 samples. *J. Photochem. Photobiol. A Chem.* **2022**, *430*, 113970. [CrossRef]
10. Simon, S.M.; George, G.; Sajna, M.S.; Prakashan, V.P.; Jose, T.A.; Vasudevan, P.; Saritha, A.C.; Biju, P.R.; Joseph, C.; Unnikrishnan, N.V. Recent advancements in multifunctional applications of sol-gel derived polymer incorporated TiO_2-ZrO_2 composite coatings: A comprehensive review. *Appl. Surf. Sci. Adv.* **2021**, *6*, 100173. [CrossRef]
11. Rafieerad, A.; Amiri, A.; Sequiera, G.L.; Yan, W.; Chen, Y.; Polycarpou, A.A.; Dhingra, S. Development of fluorine-free tantalum carbide MXene hybrid structure as a biocompatible material for supercapacitor electrodes. *Adv. Funct. Mater.* **2021**, *31*, 2100015. [CrossRef] [PubMed]
12. Kannan, K.; Radhika, D.; Nesaraj, A.S.; Sadasivuni, K.K.; Reddy, K.R.; Kasai, D.; Raghu, A.V. Photocatalytic, antibacterial and electrochemical properties of novel rare earth metal oxides-based nanohybrids. *Mater. Sci. Energy Technol.* **2020**, *3*, 853–861. [CrossRef]
13. Zhao, C.-X.; Liu, J.-N.; Wang, J.; Ren, D.; Li, B.-Q.; Zhang, Q. Recent advances of noble-metal-free bifunctional oxygen reduction and evolution electrocatalysts. *Chem. Soc. Rev.* **2021**, *50*, 7745–7778. [CrossRef]
14. Da, Y.; Li, X.; Zhong, C.; Deng, Y.; Han, X.; Hu, W. Advanced characterization techniques for identifying the key active sites of gas-involved electrocatalysts. *Adv. Funct. Mater.* **2020**, *30*, 2001704. [CrossRef]
15. Hanafi, M.F.; Sapawe, N. An overview of recent developments on semiconductor catalyst synthesis and modification used in photocatalytic reaction. *Mater. Today Proc.* **2020**, *31*, A151–A157. [CrossRef]
16. Kubiak, A.; Siwińska-Ciesielczyk, K.; Jesionowski, T. Titania-based hybrid materials with ZnO, ZrO_2 and MoS_2: A review. *Materials* **2018**, *11*, 2295. [CrossRef]
17. Li, Y.; Wang, X.; Liu, M.; Luo, H.; Deng, L.; Huang, L.; Wei, S.; Zhou, C.; Xu, Y. Molybdenum Disulfide Quantum Dots Prepared by Bipolar-Electrode Electrochemical Scissoring. *Nanomaterials* **2019**, *9*, 906. [CrossRef]
18. Sharaf, I.M.; Laifi, J.; Alraddadi, S.; Saad, M.; Koubesy, M.S.I.; Elewa, N.N.; Almohiy, H.; Ismail, Y.M.; Soldatov, A.; Aboraia, A.M. Unraveling the effect of Cu doping on the structural and morphological properties and photocatalytic activity of ZrO_2. *Heliyon* **2024**, *10*, e23848. [CrossRef]
19. Mishra, K.; Devi, N.; Siwal, S.S.; Gupta, V.K.; Thakur, V.K. Hybrid semiconductor photocatalyst nanomaterials for energy and environmental applications: Fundamentals, designing, and prospects. *Adv. Sustain. Syst.* **2023**, *7*, 2300095. [CrossRef]
20. Zhang, X.; Li, L.; Zhou, Q.; Liang, X.; Liu, D. Facile synthesis of novel gully-like double-sized mesoporous structural Sr-doped ZrO_2–TiO_2 composites with improved photocatalytic efficiency. *J. Solid State Chem.* **2019**, *269*, 375–385. [CrossRef]
21. Wang, Z.; Nayak, P.K.; Caraveo-Frescas, J.A.; Alshareef, H.N. Recent developments in p-Type oxide semiconductor materials and devices. *Adv. Mater.* **2016**, *28*, 3831–3892. [CrossRef] [PubMed]
22. Selvam, N.C.S.; Manikandan, A.; Kennedy, L.J.; Vijaya, J.J. Comparative investigation of zirconium oxide (ZrO_2) nano and microstructures for structural, optical and photocatalytic properties. *J. Colloid Interface Sci.* **2013**, *389*, 91–98. [CrossRef]
23. Gatti, T.; Lamberti, F.; Mazzaro, R.; Kriegel, I.; Schlettwein, D.; Enrichi, F.; Lago, N.; Di Maria, E.; Meneghesso, G.; Vomiero, A. Opportunities from doping of non-critical metal oxides in last generation light-conversion devices. *Adv. Energy Mater.* **2021**, *11*, 2101041. [CrossRef]
24. Zhang, Y.; Jie, W.; Chen, P.; Liu, W.; Hao, J. Ferroelectric and piezoelectric effects on the optical process in advanced materials and devices. *Adv. Mater.* **2018**, *30*, 1707007. [CrossRef]
25. Wang, W.; Tadé, M.O.; Shao, Z. Research progress of perovskite materials in photocatalysis-and photovoltaics-related energy conversion and environmental treatment. *Chem. Soc. Rev.* **2015**, *44*, 5371–5408. [CrossRef]
26. Kumari, N.; Sareen, S.; Verma, M.; Sharma, S.; Sharma, A.; Sohal, H.S.; Mehta, S.K.; Park, J.; Mutreja, V. Zirconia-based nanomaterials: Recent developments in synthesis and applications. *Nanoscale Adv.* **2022**, *4*, 4210–4236. [CrossRef]

27. Batool, S.S.; Saleem, R.; Khan, R.R.M.; Saeed, Z.; Pervaiz, M.; Summer, M. Enhancing photocatalytic performance of zirconia-based nanoparticles: A comprehensive review of factors, doping strategies, and mechanisms. *Mater. Sci. Semicond. Process.* **2024**, *178*, 108419. [CrossRef]
28. Zhang, S.; Li, B.; Wang, X.; Zhao, G.; Hu, B.; Lu, Z.; Wen, T.; Chen, J.; Wang, X. Recent developments of two-dimensional graphene-based composites in visible-light photocatalysis for eliminating persistent organic pollutants from wastewater. *Chem. Eng. J.* **2020**, *390*, 124642. [CrossRef]
29. Porcu, S.; Secci, F.; Ricci, P.C. Advances in hybrid composites for photocatalytic applications: A review. *Molecules* **2022**, *27*, 6828. [CrossRef]
30. Chatterjee, D.; Dasgupta, S. Visible light induced photocatalytic degradation of organic pollutants. *J. Photochem. Photobiol. C Photochem. Rev.* **2005**, *6*, 186–205. [CrossRef]
31. Gurushantha, K.; Anantharaju, K.S.; Renuka, L.; Sharma, S.C.; Nagaswarupa, H.P.; Prashantha, S.C.; Vidya, Y.S.; Nagabhushana, H. New green synthesized reduced graphene oxide–ZrO_2 composite as high performance photocatalyst under sunlight. *RSC Adv.* **2017**, *7*, 12690–12703. [CrossRef]
32. Li, Z.; Li, K.; Du, P.; Mehmandoust, M.; Karimi, F.; Erk, N. Carbon-based photocatalysts for hydrogen production: A review. *Chemosphere* **2022**, *308*, 135998. [CrossRef] [PubMed]
33. Osman, A.; Elhakeem, A.; Kaytbay, S.; Ahmed, A. A comprehensive review on the thermal, electrical, and mechanical properties of graphene-based multi-functional epoxy composites. *Adv. Compos. Hybrid Mater.* **2022**, *5*, 547–605. [CrossRef]
34. Sambyal, S.; Sharma, R.; Mandyal, P.; Balou, S.; Gholami, P.; Fang, B.; Shandilya, P.; Priye, A. Advancement in two-dimensional carbonaceous nanomaterials for photocatalytic water detoxification and energy conversion. *J. Environ. Chem. Eng.* **2023**, *11*, 109517. [CrossRef]
35. Liu, X.; Li, S.; Yu, J.; Zhu, Y.; Lin, K.; Wang, B.; Cai, R.; Ekren, D.; Lewis, D.; Kinloch, I.A. Enhancing the thermoelectric properties of TiO_2-based ceramics through addition of carbon black and graphene oxide. *Carbon* **2024**, *216*, 118509. [CrossRef]
36. Purohit, S.; Yadav, K.L.; Satapathi, S. Metal halide perovskite heterojunction for photocatalytic hydrogen generation: Progress and future opportunities. *Adv. Mater. Interfaces* **2022**, *9*, 2200058. [CrossRef]
37. Singh, B.R.; Shoeb, M.; Khan, W.; Naqvi, A.H. Synthesis of graphene/zirconium oxide nanocomposite photocatalyst for the removal of rhodamineB dye from aqueous environment. *J. Alloys Compd.* **2015**, *651*, 598–607. [CrossRef]
38. Sikdar, S.; Banu, A.; Ali, S.; Barman, S.; Kalar, P.L.; Das, R. Micro-structural Analysis and Photocatalytic Properties of Green Synthesized t-ZrO_2 Nanoparticles. *ChemistrySelect* **2022**, *7*, e202103953. [CrossRef]
39. Shinde, H.M.; Bhosale, T.T.; Gavade, N.L.; Babar, S.B.; Kamble, R.J.; Shirke, B.S.; Garadkar, K.M. Biosynthesis of ZrO_2 nanoparticles from Ficus benghalensis leaf extract for photocatalytic activity. *J. Mater. Sci. Mater. Electron.* **2018**, *29*, 14055–14064. [CrossRef]
40. Dhandapani, C.; Narayanasamy, R.; Karthick, S.N.; Hemalatha, K.V.; Selvam, S.; Hemalatha, P.; Kumar, M.S.; Kirupha, S.D.; Kim, H.J. Drastic photocatalytic degradation of methylene blue dye by neodymium doped zirconium oxide as photocatalyst under visible light irradiation. *Optik* **2016**, *127*, 10288–10296. [CrossRef]
41. Aljawrneh, B.; Ocak, Y.S.; Albiss, B.A.; Dwiri, A.; Tawalbeh, M.; Al-Othman, A. ZrO_2 nanoparticles for effective dye degradation in wastewater: Synthesis, characterization, and photocatalytic performance under sunlight. *J. Alloys Compd.* **2024**, *1008*, 176522. [CrossRef]

catalysts

MDPI

Article

Unlocking Synergistic Catalysis in NiP: Dual Role of Electronic Structure and Lewis Acidity for Enhanced Oxygen Evolution Reaction

Jiazhou Liang [1], Jiawei Li [1], Jiani Yan [1], Andrew M. Rappe [2,*] and Jing Yang [1,*]

1 School of Chemical Engineering and Technology, Sun Yat-sen University, Zhuhai 519082, China
2 Department of Chemistry, University of Pennsylvania, Philadelphia, PA 19104-6323, USA
* Correspondence: rappe@sas.upenn.edu (A.M.R.); yangjing25@mail.sysu.edu.cn (J.Y.)

Abstract: Nickel phosphides (Ni_xP_y) are recognized as promising alternatives to noble-metal catalysts for the oxygen evolution reaction (OER). NiP, consisting of the equal stoichiometric ratio of Ni and P, could help quantify the catalytic effect of P and Ni. In this work, density functional theory (DFT) is employed to investigate the OER mechanism on NiP surfaces. We found that P atoms help stabilize O* at the adsorption sites. The rich electron donation from the Ni atom can alter the local charge distribution and enhance the interaction between O* and P atoms. Both oxygen intermediate adsorption energy and OER overpotential exhibit linear correlations with the charge of adsorption sites. Electron loss at the site induces the overall system to exhibit Lewis acidic characteristics, facilitating the OER and leading to a substantial overpotential reduction of up to 0.61 V compared to Lewis basic structures. Leveraging electronic structure theory and Lewis acid–base theory, we offer a new insight into the OER mechanism on the NiP surface, demonstrating that the catalytic activity of bulk metallic surface materials like NiP can be optimized by tailoring the local surface chemical environment.

Keywords: NiP; oxygen evolution reaction; surface site charge; surface adsorption; density functional theory

1. Introduction

The global surge in energy demand, coupled with the depletion of fossil fuel reserves and their severe environmental impact, has intensified the urgency for researchers to explore green energy alternatives and develop more efficient sustainable energy conversion technologies [1–3]. Hydrogen is widely recognized as a clean and renewable energy carrier due to its high energy density and zero carbon emissions. The most promising hydrogen production method thus far is water electrocatalytic decomposition [4–7]. In the water-splitting reaction, the OER is a critical bottleneck, as it entails a complex four-electron process that is often regarded as the rate-limiting step of the entire catalytic cycle. Consequently, the development of efficient OER electrocatalysts is of paramount importance [8–10].

Precious metals and their oxides, particularly Ru, Ir, RuO_2, and IrO_2, demonstrate superior OER electrocatalytic performance. However, their widespread adoption is severely limited by their prohibitive cost. For instance, in highly efficient acidic proton exchange membrane water electrolyzers (PEMWEs), recognized as a premier hydrogen production technology, anode OER electrocatalysts are still limited to precious Ir and Ru oxides [11].

https://doi.org/10.3390/catal15050457

Developing highly active and noble-metal-free electrocatalysts is critical for the large-scale application of PEMWEs [12–15]. Earth-abundant transition-metal catalysts have garnered significant attention in recent years for their potential in electrocatalytic water splitting [16–18]. Nickel phosphides, a representative transition-metal phosphide, have emerged as a notable contender. The favorable combination of their low preparation costs and high catalytic activity renders them promising candidates for OER catalysis [19–21]. It has been reported that the dissolution step in pulsed reverse electrodeposition (PRED) regulates the chemical state of phosphorus in NiP films, where increased dissolution promotes the formation of removable phosphates to enhance the OER activity of NiP films, achieving an average reduction of 55.1 mV in overpotential at a current density of 10 mA/cm^2 compared to constant potential deposition (CPD) results; XPS spectra reveal that the removal of phosphates exposes more active nickel oxide sites, thereby improving OER activity [22]. Notably, Ni_xP_y with a relatively high metal content exhibits superior activity towards OER, as exemplified by $Ni_{12}P_5$, whose OER performance surpasses that of Ni_2P due to its higher Ni content facilitating the formation of a more catalytically active $NiOOH/Ni(OH)_2$ shell layer on the surface [23–29]. Phosphorus within NiP functions as an excellent supporting layer beneath the surface of nickel oxide or oxyhydroxide during the OER, enhancing the stability of the catalyst [29,30]. In addition, under actual reaction conditions, solid surfaces undergo geometric, atomic, and electronic reconstructions to minimize surface free energy. The reconstructed atomic surface stoichiometry may either mirror the bulk termination stoichiometry or differ from it [31–33]. Under practical conditions, nickel phosphides have been proven to undergo various surface reconstructions, including the formation of oxide layers, which significantly alter their catalytic performance compared to the bulk-terminated surfaces [34,35]. However, in our case of exploring the OER mechanism with the material stoichiometric ratio effect, we will focus more on the primitive surface and figure out the actual role of P element in the material. Therefore, it is essential to systematically investigate the NiP surface structural reconstructions and thoroughly explore the OER mechanism. Notably, transition-metal phosphides, including NiP, offer abundant metal and nonmetal adsorption sites for OER intermediates, introducing additional complexity beyond traditional metal-based systems. Most researchers believe that conventional descriptors, such as the d-band model-related features, prove insufficient in providing comprehensive insights into these complex systems [36–38]. Therefore, a systematic analysis of the OER reaction mechanism at different active sites is crucial for the rational design of efficient OER catalysts.

In this work, we investigate the surface reconstruction and OER activity of NiP(100), (010), and (001) using DFT. We identified the most stable surfaces among the various surface reconstructions of those low index surfaces. During the investigation of OER activity, we found that the charges at the adsorption sites are linearly correlated with the oxygen adsorption intermediates and even the OER overpotential. The results show that the charge transfer at the site can regulate the adsorption energy of the intermediate and impact the performance of the catalysts. We hope that our results can provide a reference for the design of base-metal catalysts.

2. Results and Discussions

2.1. Stability of NiP Surfaces

To identify the most stable surface, we computed the surface energies of various NiP slab models with different surface terminations. Here, we have studied the surface stability of (100)-Pris: P_2, (100)-C1:Ni_4P_2, (100)-C2:Ni_2P_2, (001)-Pris: Ni_2P_2, (001)-C1:Ni_2P_2, (010)-Pris: Ni_2P_2, (010)-C1:P_4, and (010)-C2:P_2, where the symbol is defined as (facet)-Reconstructions: termination (Pris represents the pristine bulk structure termination, C1 represents the cutting of one layer of the top surface atoms, C2 represents the cutting of two layers of the top surface atoms).

Based on our calculations, we have found that P-rich surface terminations are the most energetically favorable for NiP among the low-index surfaces, which is consistent with previous theoretical and experimental studies [39–42]. The surface energy diagram of the slabs of all three lattice planes showed that the surfaces with P atom terminations have low surface energies, indicating that they are thermodynamically stable. Before the formation of bulk NiP region (-1.2 eV $< \Delta\mu_P <$ 0 eV), four stable surface terminations with lower surface energy ((010)-C1:P_4, (010)-C2:P_2, (001)-C1:Ni_2P_2, and (100)-C1:Ni_4P_2) are shown in Figure 1b. Among these surfaces, NiP(010) exhibits the lowest surface energy, indicating the most stable surface termination.

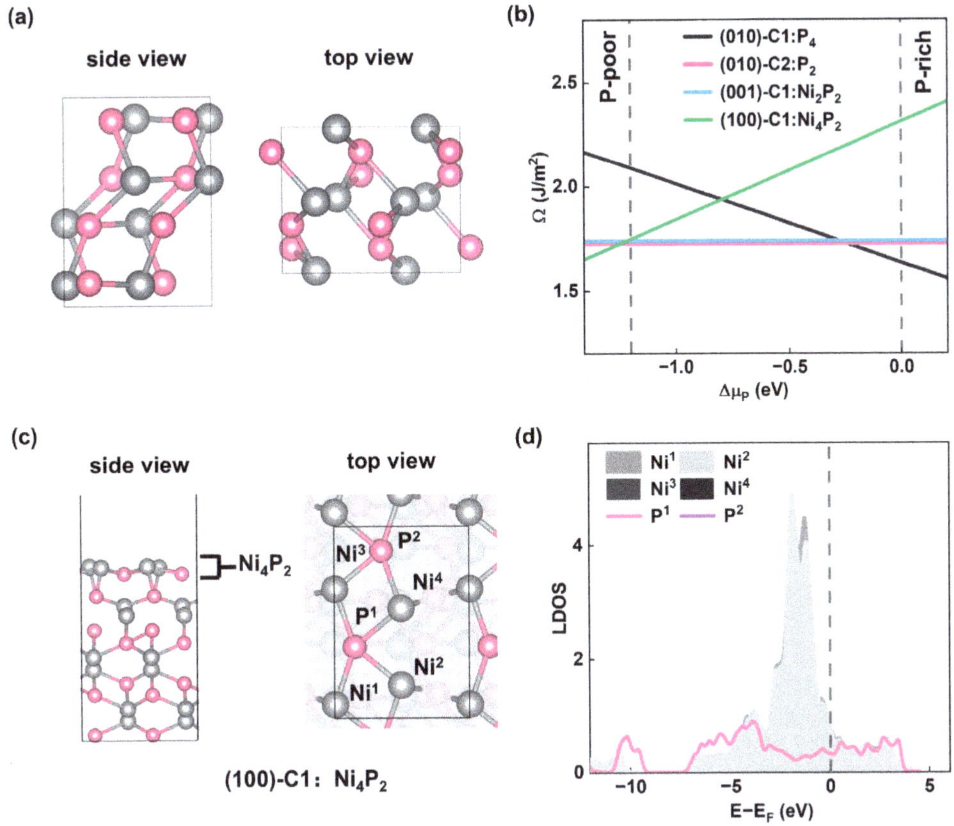

Figure 1. (**a**) Ball-and-stick diagram of the crystal structure of NiP. (**b**) Surface energies of surfaces (010)-C1:P_4, (010)-C2:P_2, (001)-C1:Ni_2P_2, and (100)-C1:Ni_4P_2 as a function of P chemical potential ($\Delta\mu_P$). (**c**) Ball-and-stick representation of (100)-C1: Ni_4P_2 surface termination atoms distribution. (**d**) DOS analyses of (100)-C1: Ni_4P_2 surface termination atoms. In the ball-and-stick images, Ni atoms are light gray, and P atoms are pink.

2.2. Stability of the Adsorption Structures

To investigate the relationship between the catalytic activity and surface active sites, we constructed multiple adsorption systems of the OER intermediates on the various adsorption sites of the surfaces. Judging by the surface density of states (DOSs), we found that many surface sites are similar in terms of geometric position and electronic structure.

Taking (100)-C1:Ni_4P_2 as an example, the surface layer consists of four Ni atoms and two P atoms (Figure 1c). Ni^1 and Ni^3 are completely equivalent due to the full overlap of the DOS (see Figure 1d). The same applies to Ni^2 and Ni^4, as well as P^1 and P^2. For the NiP(001)-C1: Ni_2P_2 surface, both the two Ni and P sites are equivalent, leading to one set of Ni-P sites. For (010)-C1:P_4 and (010)-C2:P_2 surfaces, the topmost layer only contains P atoms, and they are all equivalent within the surfaces correspondingly (Figure S5).

Given the equivalency of the surface sites, we considered a one monolayer (1 ML) adsorption condition of the OER cycle species. Due to the simple structure of the adsorbates, such as O*, HO*, and HOO*, we chose the on-top adsorption site for each case. To determine stable adsorption structures, we tested four adsorption distances: 1.3 Å, 1.5 Å, 2 Å, and 2.5 Å. Our findings reveal that the diverse adsorption sites on the surfaces create a complicated atomic interaction for adsorbed OER intermediates. Notably, the available adsorption sites encompass Ni_2P-hollow, Ni on-top, P on-top, as well as the bridge sites between Ni-Ni and Ni-P atoms (see the adsorption energy (E_{ads}) for various sites in Table S3). The stable oxygen intermediate adsorption sites are Ni_2P-hollow, Ni on-top, P on-top, and the Ni-Ni and Ni-P bridge sites. For the HO* group, the stable sites are P on-top and the Ni-Ni bridge sites. Similar to the O* group, the HOO* can be stably adsorbed on Ni on-top, P on-top, and the Ni-Ni bridge sites.

2.3. OER Pathways on NiP Surfaces

To understand the mechanism of OER on the NiP surface, we investigated the OER process on the four stable NiP surfaces, as determined in the previous section (Figure 1b). We found that on the four stable surfaces, there are two reaction pathways for the OER, which are described as the single-site pathway and the dual-site pathway. The single-site refers to all the intermediate species adsorbing on the same site, whereas the dual-site pathway introduces a second adsorption site of the oxygen intermediate during the reaction cycle (Figure 2). A simplified version of the proposed OER mechanism as the reference pathway is provided in Figure S7. The single-site pathway occurs on all four stable surfaces, whereas the dual-site reaction can happen only on Ni_4P_2 and P_2 surfaces. For each intermediate, the preferred adsorption location can be determined by comparing the adsorption energy at various adsorption sites. For the Ni_4P_2 surface, the preferred adsorption site for HO* and HOO* is the Ni-Ni bridge. The energetically favored adsorption site for O* is Ni_2P-hollow instead of Ni-Ni bridge, which converts the pathway from single-site to dual-site (Figure 2a). The adsorption energy difference between O* adsorption on Ni-Ni bridge and Ni_2P-hollow is 1.84 eV, according to our calculation. The dual-site pathway usually involves sublayer atoms and includes both P and Ni atoms. For the P_2 surface, both HO* and HOO* are stably adsorbed on the P atom top site. O* is preferentially adsorbed on a Ni-P bridge site, forming another dual-site pathway with the adsorption energy 0.42 eV lower than the O* adsorbed onto the P atom (Figure 2b). For the P_4 and Ni_2P_2 surface, all the OER intermediates adsorb more strongly onto the P on-top site than onto other sites, which can be described as a single-site pathway (Figure S6). A similar dual-site pathway can be found in the NiN_4C_4 moiety, where HO* and O* preferentially adsorb onto the C atom, whereas HOO* tends to reside on the central Ni atom, as demonstrated by Fei and coworkers [43]. However, in our dual-site case, HO* and HOO* are both adsorbed onto metal or non-metal sites (Ni-Ni bridge and P atom), while O* prefers the P-containing site (Ni_2P-hollow and Ni-P bridge). We subsequently evaluated the OER catalytic activity of various sites based on the established reaction paths.

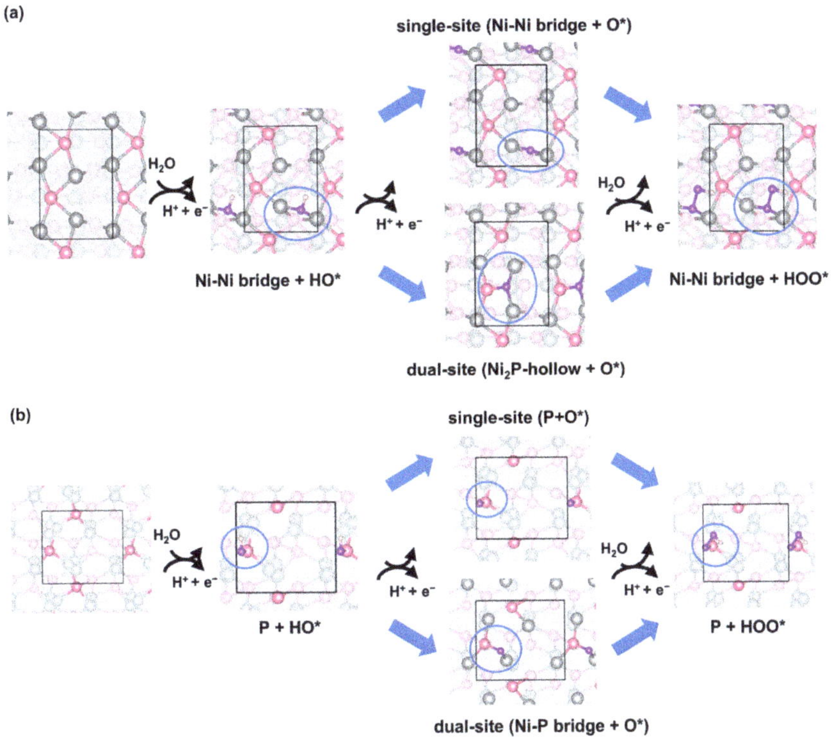

Figure 2. Proposed mechanism of the OER as our reference pathway. The single-site and dual-site pathways towards OER of (**a**) Ni_4P_2 surface and (**b**) P_2 surface. In the ball-and-stick images, Ni atoms are light gray, P atoms are pink, O atoms are purple, and H atoms are white.

2.4. Investigation of Electrocatalytic OER Mechanism

To investigate the relationship between the nature of surface active sites and catalytic activity, we calculated the Gibbs free energy of the OER intermediates at $U = 0$ V and $U = 1.23$ V according to the proposed reaction pathways (Figure 3a–c). One important parameter used to evaluate catalytic activity is the OER overpotential, which is determined by the free energy of the rate-determining step (RDS). For the single-site OER pathway of P_4 and Ni_2P_2 surfaces, the RDS is O* oxidation to HOO* with overpotentials as large as 1.54 V and 1.57 V ($U = 0$ V) (Figure 3c). Moreover, the adsorption energies of the O* on P_4 and Ni_2P_2 surfaces are very similar (Figure 3d), which leads to the overpotential being pretty similar to each other. For the P_2 surface, the RDS is the oxidation of O* to HOO* for both the single-site and dual-site reaction pathways (Figure 3b). Notably, the overpotential for the single-site pathway is 1.68 V, whereas the overpotential is significantly lower, at 1.07 V, for the dual-site pathway. Therefore, the OER prefers the dual-site pathway on the P_2 surface. Furthermore, the involvement of the Ni atom enhances the adsorption of O* on the P_2 surface ($E_{ads}(O*) = -4.56$ eV) compared to the single-site pathway with the P atom-only interaction. However, the joint interaction between Ni and P can cause too strong adsorption of the O*, hindering the OER cycle. In the case of the Ni_4P_2 dual-site pathway, O* adsorbs onto the Ni_2P-hollow site with a strong adsorption energy of -5.16 eV and an overpotential of 1.30 V, hindering the participation of O* in subsequent reactions. By contrast, Ni_4P_2 with the single-site pathway shows the lowest overpotential of 0.78 V amongst all the surfaces, featuring the oxidation of HOO* to O_2 as the RDS (Figure 3a) rather than the more common RDS of O* to HOO*. The metal atom site

(Ni-Ni bridge site) provides moderate adsorption to the intermediates and activates the reaction cycle (Figure 3d). According to the Sabatier principle, the adsorption energy of the intermediate needs to fall within an appropriate range to ensure the equilibrium between the adsorption and desorption of the intermediate [44]. Given that O* exhibits a strong adsorption affinity to the Ni_2P-hollow site, more energy is required to desorb O*, resulting in a larger overpotential and shifting the RDS to the oxidation of O* to HOO*; therefore, it hinders subsequent reactions. Consequently, for the Ni_4P_2 surface, the OER prefers a single-site pathway (Figure 2a) rather than a dual-site one.

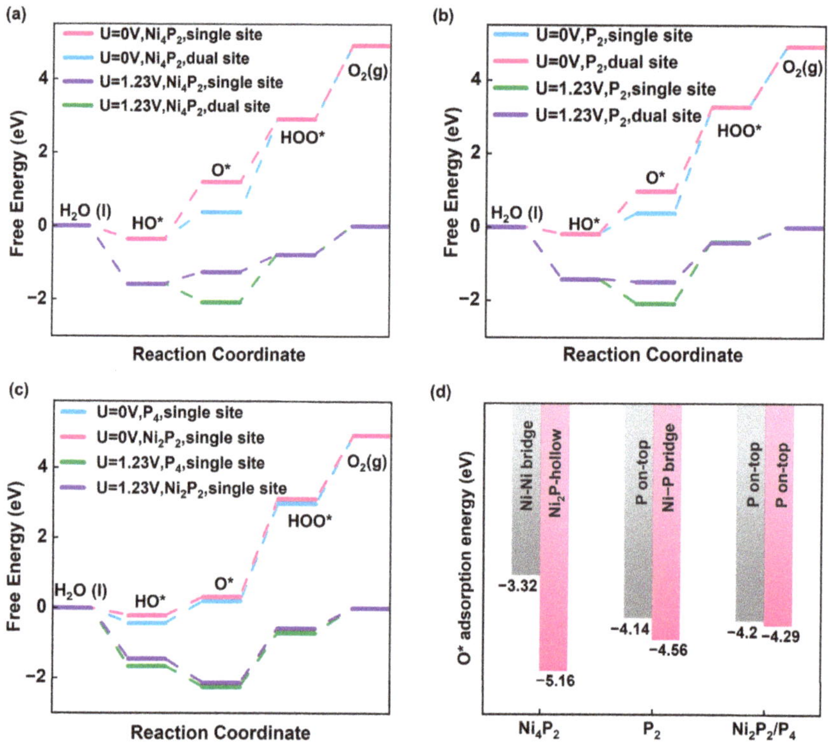

Figure 3. Free energy diagrams of OER for (**a**) Ni_4P_2 and (**b**) P_2 with single-site and dual-site pathways, and (**c**) Ni_2P_2 and P_4 with single-site pathway at $U = 0$ V and 1.23 V. (**d**) O* adsorption energy for the available adsorption sites.

While energetic analysis is helpful for explaining the OER catalytic activity at various active sites, we present an analysis grounded in electronic structure theory to elucidate the relationship between the catalytic activity and the adsorption site characteristics.

Focusing on the RDS investigations, we found that the interaction between O* and the sites encompassing Ni and P atoms is usually stronger than in other cases. In the case of the P_2 surface, the differential charge density depicts that O* adsorption reveals a significant redistribution of charge between O* and various surface sites (Figure 4a–d). The adsorbed O* induces electron accumulation at both the Ni-P bridge site and the P site. Due to the high electronegativity of oxygen, O* tends to obtain electrons from Ni and P atoms within the adsorption sites. In the single-site pathway, O* adsorbs onto the P site and obtains electrons from the P atoms in the adsorption site. In the dual-site pathway, O* interacts with the Ni-P bridge site, and the electronic interaction between the O* and the P atoms on the Ni-P bridge site is stronger than that on the P site. Meanwhile, O* exhibits a larger charge accumulation at the Ni-P bridge site compared to that at the P site. All these phenomena can be attributed to the additional electron donation from the Ni atom, which alters the local charge distribution and enhances the interaction between O* and P atoms. The electronic interaction of O* on the Ni_4P_2 surface is similar to that on the P_2 surface (Figure S8). In the single-site pathway, strong charge accumulation and depletion can be observed between O* and Ni atoms at the Ni-Ni bridge site. In the dual-site pathway, O* acquires electrons from both the Ni atoms and the P atom residing within the Ni_2P-hollow site. Notably, on the Ni_2P-hollow site, a relatively stronger electron dissipation occurs between O* and P atoms as compared to that between Ni atoms. The strong electronic interaction between O* and P engenders a more negative adsorption energy of O* at the Ni_2P site ($E_{ads}(\mathrm{O}*) = -5.16$ eV) in comparison to that of the Ni-Ni bridge site composed solely of Ni atoms ($E_{ads}(\mathrm{O}*) = -3.32$ eV), thereby conferring enhanced adsorption stability.

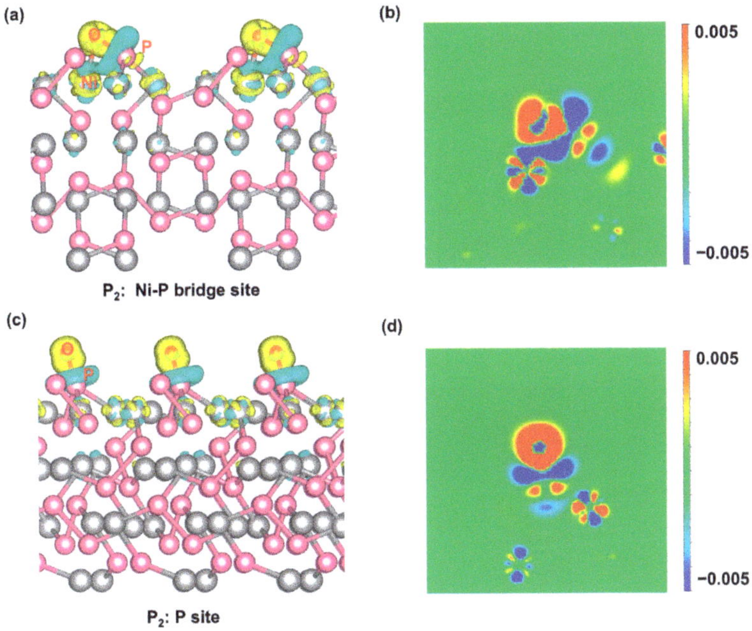

Figure 4. Differential charge density distribution of O* adsorbed onto the (**a**) Ni-P bridge site and (**c**) P site of the P_2 surface (isosurface value = 0.007 eÅ^{-3}). The yellow regions represent electron charge accumulation, whereas the blue regions indicate charge depletion. (**b**,**d**) Cross-sectional views of the differential charge density in Figure 4a,c, respectively. The scale bar represents the magnitude of the electronic charge density difference; blue represents electronic charge depletion, and red represents electronic charge accumulation.

Thoroughly analyzing the density of states (DOSs) of the various adsorption structures among different surface terminations, we found that P atoms tend to assist in stabilizing O* at the adsorption sites. When Ni and P combine at the site, such a P involvement can further activate the Ni sites in the catalytic cycle, driving Ni as an electron donor. The interaction of O* at P sites of various surfaces presents a similar behavior in that the major oxygen p orbital peak lies around −5 eV and overlaps with the phosphorus p orbital, indicating a strong and stable chemical bond formation between O* and P. The non-bonding oxygen p orbital peak can be found around −2.5 eV with an almost half-full occupation, making it difficult for the O* to propagate further reactions (Figure 5b,d). This phenomenon is attributed to relatively low adsorption energy and high overpotential compared to other cases. When O* adsorbs onto the Ni-Ni bridge site found on the Ni_4P_2 surface (Figure 5a), the oxygen p orbital shifts to the higher energy region in general compared to O*-P case (Figure 5b), and the previous non-bonding part hybridizes with the Ni d orbital crossing the Fermi level, forming an active state to proceed the further reaction along the OER cycle. When a P atom presents within the adsorption site, as is the case for the Ni-P bridge site (Figure 5c), the major p contribution of the O* falls back to around -5 eV, and the reduced overlap between the oxygen p orbital and Ni d orbital suggests a weakened O*-Ni interaction. However, the d band center of Ni shifts higher to be close to the Fermi level, which activates the Ni atom in the Ni-P bridge site for the next reaction step. In this case, the Ni atom provides its excess d electrons to O* (Figure 4a), leaving the unsaturated site for the opportunity of a second water molecule to break to form HOO* (SI Equation (S3)). Hence, the dual-site pathway is preferred on the P_2 surface.

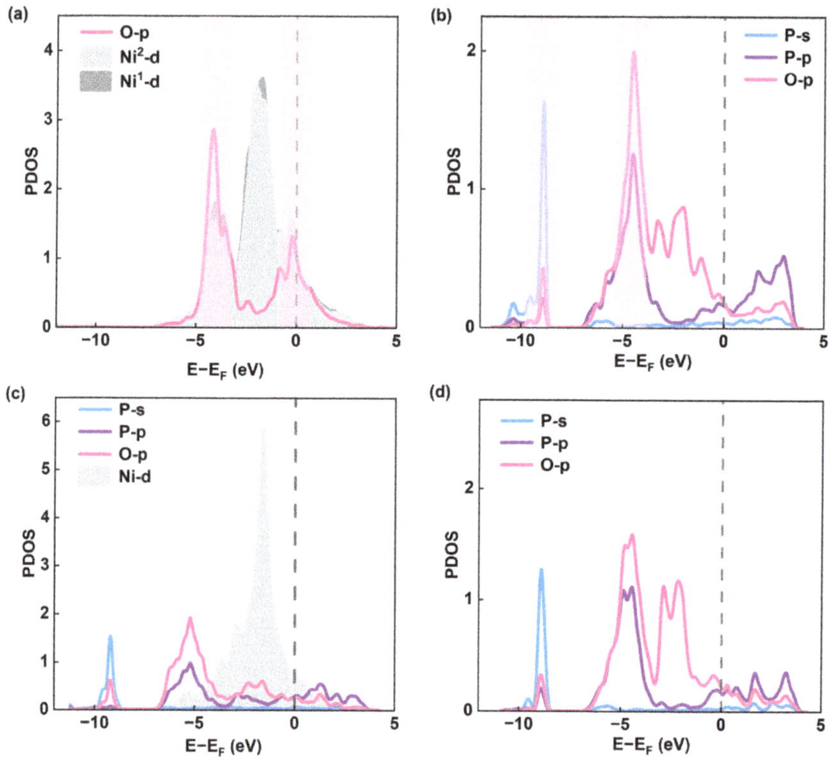

Figure 5. PDOS analyses of O* adsorbed onto (**a**) Ni-Ni bridge site of the Ni_4P_2 surface and (**b**) P site of the P_4 surface. PDOS analyses of O* adsorbed onto (**c**) Ni-P bridge site in the dual-site pathway and (**d**) P site in the single-site pathway of the P_2 surface.

Except for the Ni_4P_2 surface, the RDS of the OER reaction on P_4, P_2, and Ni_2P_2 consistently involves the oxidation of O* to HOO*. Therefore, we performed Bader charge analysis on adsorption sites and oxygen intermediates to investigate the charge transfer (Table S4). Previous studies have demonstrated that substantial charge transfer between active sites and adsorbates typically correlates with lower adsorption energy [45–48]. To investigate the potential mechanism of the RDS on various NiP surfaces, we calculated the adsorption energy of O* and the amount of charge transferred at the adsorption site. We observed a strong linear correlation between the amount of charge transferred at the adsorption site and the adsorption energy of O* (Figure 6a). Within the charge range of

1 to 2 $|e|$ for the adsorption site, the adsorption energy of O* exhibits a decreasing trend with increasing charge, indicating that charge transfer occurring at the adsorption sites can effectively modulate the adsorption energy of O*, which provides insights for optimizing intermediate adsorption to regulate the performance of OER catalysts [49,50]. In addition, we calculated the Bader charge of the sites before adsorption. We found that the charge of the adsorption site and the OER overpotential are linearly related. The overpotential decreases as the charge lost from the adsorption site increases (Figure 6b). It is accepted that the adsorption properties of reaction intermediates are closely linked to the catalytic activity of the reaction [51,52]. Thus, the binding energy differences of intermediates ($\Delta G_{O*} - \Delta G_{HO*}$) are widely used to evaluate the OER activity. However, directly regulating the value of $\Delta G_{O*} - \Delta G_{HO*}$ to improve catalytic performance is impractical. Our discovery establishes a direct linear relationship between the charge of the adsorption site and OER overpotential, indicating that catalytic performance can be directly optimized by regulating the charge of the adsorption site, thus providing a more precise electronic regulation strategy for catalyst design. Moreover, we observed that the adsorption sites containing Ni atoms tend to form a Lewis acid system (Figure 6b). Previous studies have shown that OER intermediates exhibit behaviors similar to Lewis bases with lone electron pairs [53–55]. When the adsorption site loses electronic charge, the electron cloud density around the site atoms decreases relatively, enhancing its electron-accepting ability, making the overall structure exhibit Lewis acid characteristics, which is conducive to the formation and adsorption of oxygen-containing intermediates, thereby promoting the progress of OER. In this study, compared to structures with Lewis basic characteristics, those exhibiting Lewis acidity show a remarkable decrease in overpotential by up to 0.61 V. Moreover, the charge of the adsorption sites can directly influence the performance of the NiP catalysts, indicating that even though NiP exhibits metallic properties, the local chemical environment can still determine the catalytic activity. In other words, the catalytic activity of bulk metallic surface material, such as NiP, can be optimized by tailoring the local chemical environment on the surface.

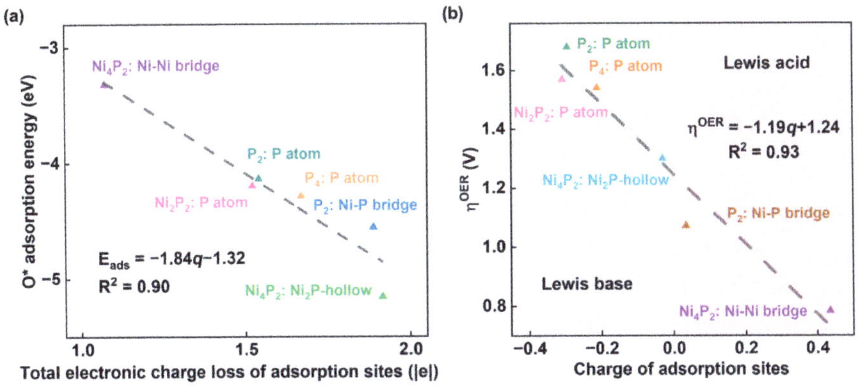

Figure 6. (**a**) The relationship between adsorption energy (O*) and Bader charge of the adsorption atoms (Ni and P). (**b**) The relationship between the charge of the sites before adsorption and OER overpotential η^{OER}. The gray dashed lines are the linearly fitted lines.

3. Computational Methods

For the DFT calculations, we use the Quantum ESPRESSO software package (version 7.1) with a plane-wave basis set [56,57]. The exchange-correlation energy was calculated using the Perdew–Burke–Ernzerhof (PBE) formulation of the generalized gradient approximation (GGA) [58]. Non-local norm-conserving pseudopotentials constructed with the OPIUM code are used [59,60]. The van der Waals interaction (vdW) was approximated using the DFT-D3 method [61]. During NiP slab and the adsorption structure optimization, only the top half of the atoms are relaxed, keeping the lower half fixed. The convergence accuracy of the total energy of the system is 10^{-5} eV, and the convergence accuracy of the forces is set to 10^{-4} eV/Å. The k-point sampling in the first Brillouin zone for the supercells is shown in Table S2.

The NiP cell structure in the international crystal structure database (ICSD; collection code 27159, space group Pbca) is selected as the bulk structure for NiP calculation [62]. The lattice parameters of completely relaxed NiP are a = 6.103 Å, b = 4.929 Å, and c = 6.905 Å, which are within 1% of the experimental values (a = 6.050 Å, b = 4.881 Å, and c = 6.890 Å). We construct a slab model to simulate the surface structures. The $(2 \times 1 \times 1)$ supercell with two formula units of NiP along the z-direction ensures the decoupling of the continuous plates. Adsorption was allowed on one side of the slab, and to ensure the decoupling of the consecutive slabs, a 15 Å-thick vacuum region is employed, with an artificial dipole correction inserted into the center of the vacuum layer [63]. This work studies three surface directions of NiP: (001), (010), and (100), and uses a fully relaxed unit cell to construct the supercell structure to construct the slab model. Interface tailoring can effectively improve the activity and stability of the catalyst [64–66]. Judging by the pristine surface slab geometry directly generated from the bulk, the slab layer geometry repeats itself within two layers. Therefore, our search for low-energy reconstructions starts with one or the other layer of the original crystal structure to ensure that the search covers the possible reconstructions (see the Supporting Information (SI) for more details).

The catalytic activity of OER catalysts depends on the Gibbs free energies of intermediates (ΔG_{HO*}, ΔG_{O*}, and ΔG_{HOO*}) [67]. For an ideal OER cycle, the Gibbs free energy change between two adjacent intermediates should be 1.23 eV (U = 0 V). By calculating the free energy of each OER reaction step, we can use the calculated OER overpotential to evaluate the catalyst's OER performance. More computational details are provided in the Supporting Information.

4. Conclusions

In summary, we systematically investigated the OER mechanism on NiP surfaces using DFT calculations. Electronic structure theory analysis reveals that the surface P atoms work as stabilizers to stabilize O* at the adsorption sites. The excessive electrons from Ni atoms can alter the local charge distribution and enhance the interaction between O* and P atoms. Moreover, we found that the charge of the adsorption sites can be linearly correlated with both the adsorption energy of oxygen intermediates and the OER overpotential. When the site undergoes charge loss, the entire system demonstrates Lewis acid characteristics, making it favorable for OER. Compared to structures with Lewis basic characteristics, those exhibiting Lewis acidity show a remarkable decrease in overpotential by up to 0.61 V. This work elucidates the OER mechanism on the NiP surface, demonstrating that the catalytic activity of bulk metallic surface materials like NiP can be optimized by tailoring the local chemical environment through strategies such as doping-induced Lewis acid site engineering. These findings establish a valuable framework for guiding base metal catalyst design. Future work could explore doping strategies to introduce Lewis acidic sites and investigate surface oxidative reconstruction under high anodic potentials and low-pH conditions to better evaluate and enhance catalyst performance.

Supplementary Materials: The following supporting information can be downloaded at: https://www.mdpi.com/article/10.3390/catal15050457/s1, Figure S1: (a) (100)-Pris; (b) (100)-C1; (c) (100)-C2. Pris represents the original surface, C1 represents the cutting of one layer of surface termination atoms, and C2 represents the cutting of two layers of surface termination atoms. The pink sphere represents P atoms and the gray sphere represents Ni atoms. Figure S2: Side view of NiP(100) terminations. (a) (100)-Pris terminations P_2 (b) (100)-C1 terminations Ni_4P_2. (c) (100)-C2 terminations Ni_2P_2. (d) Surface energy of different terminations. Figure S3: Side view of NiP(010) terminations. (a) (010)-Pris terminations Ni_2P_2. (b) (010)-C1 terminations P_4. (c) (010)-C2 terminations P_2. (d) Surface energy of different terminations. Figure S4: Side view of NiP(001) terminations. (a) (001)-Pris terminations Ni_2P_2. (b) (001)-C1 terminations Ni_2P_2. (c) Surface energy of different terminations. Figure S5: surface termination atoms distribution of (a) (001)-C1: Ni_2P_2, (c) (010)-C2: P_2, and (e) (010)-C1: P_4. DOS analyses of (b) (001)-C1: Ni_2P_2, (d) (010)-C2:P_2, and (f) (010)-C1: P_4 surface atoms. Figure S6: The single-site pathway towards OER of (a) Ni_2P_2 surface and (b) P_4 surface. In the ball-and-stick images, Ni atoms are in light gray, P atoms are in pink, O atoms are in purple, and H atoms are in white. Figure S7: (a) Surface site schematic. (b) A schematic depiction of single-site (L) and dual-site (R) mechanism. Figure S8: Differential charge density distribution of O* adsorbed onto the (a) Ni-Ni bridge site and (c) Ni_2P-hollow site of the Ni_4P_2 surface (Isosurface value = 0.007 $eÅ^{-3}$). (b) and (d) Cross-sectional views of the differential charge density in Figure S8a and S8c, respectively. The scale bar represents the magnitude of the charge density difference; the blue ends stand for charge depletion, whereas the red ends represent charge accumulation. Figure S9: Differential charge density distribution of O* adsorbed onto the P site of the (a) Ni_2P_2 and (b) P_4 surface. Table S1: Magnetic moments of the system. Table S2: k-points setting for different structures. Table S3: O*, HO*, and HOO* adsorption energies for the available adsorption sites. Table S4: Calculated Bader charge of the adsorption systems formed by O* at different adsorption sites. Refs. [68,69].

Author Contributions: J.L. (Jiazhou Liang): investigation, conceptualization, methodology, validation, and writing—original draft; J.L. (Jiawei Li): investigation; J.Y. (Jiani Yan): data curation; A.M.R.: review and editing; J.Y. (Jing Yang): conceptualization, supervision, writing—review and editing, and funding acquisition. All authors have read and agreed to the published version of the manuscript.

Funding: This research is funded by the National Key R&D Program of China (No. 2023YFB2405800).

Data Availability Statement: The data that support the findings of this study are available upon request.

Acknowledgments: The authors acknowledge the Hefei Advanced Computing Center for providing high-performance computing resources.

Conflicts of Interest: The authors declare no conflicts of interest.

References

1. Chow, J.; Kopp, R.J.; Portney, P.R. Energy resources and global development. *Science* **2003**, *302*, 1528–1531. [CrossRef]
2. Chu, S.; Majumdar, A. Opportunities and challenges for a sustainable energy future. *Nature* **2012**, *488*, 294–303. [CrossRef]
3. Shindell, D.; Smith, C.J. Climate and air-quality benefits of a realistic phase-out of fossil fuels. *Nature* **2019**, *573*, 408–411. [CrossRef]
4. Dresselhaus, M.S.; Thomas, I.L. Alternative energy technologies. *Nature* **2001**, *414*, 332–337. [CrossRef]
5. Rausch, B.; Symes, M.D.; Chisholm, G.; Cronin, L. Decoupled catalytic hydrogen evolution from a molecular metal oxide redox mediator in water splitting. *Science* **2014**, *345*, 1326–1330. [CrossRef]
6. Gao, M.-R.; Liang, J.-X.; Zheng, Y.-R.; Xu, Y.-F.; Jiang, J.; Gao, Q.; Li, J.; Yu, S.-H. An efficient molybdenum disulfide/cobalt diselenide hybrid catalyst for electrochemical hydrogen generation. *Nat. Commun.* **2015**, *6*, 5982. [CrossRef]
7. Castelvecchi, D. How the hydrogen revolution can help save the planet—And how it can't. *Nature* **2022**, *611*, 440–443. [CrossRef]
8. Xue, Z.; Li, X.; Liu, Q.; Cai, M.; Liu, K.; Liu, M.; Ke, Z.; Liu, X.; Li, G. Interfacial Electronic Structure Modulation of NiTe Nanoarrays with NiS Nanodots Facilitates Electrocatalytic Oxygen Evolution. *Adv. Mater.* **2019**, *31*, e1900430. [CrossRef]
9. Nong, H.N.; Falling, L.J.; Bergmann, A.; Klingenhof, M.; Tran, H.P.; Spöri, C.; Mom, R.; Timoshenko, J.; Zichittella, G.; Knop-Gericke, A.; et al. Key role of chemistry versus bias in electrocatalytic oxygen evolution. *Nature* **2020**, *587*, 408–413. [CrossRef]
10. Li, L.; Wang, P.; Shao, Q.; Huang, X. Recent Progress in Advanced Electrocatalyst Design for Acidic Oxygen Evolution Reaction. *Adv. Mater.* **2021**, *33*, e2004243. [CrossRef]
11. Gao, J.; Tao, H.; Liu, B. Progress of Nonprecious-Metal-Based Electrocatalysts for Oxygen Evolution in Acidic Media. *Adv. Mater.* **2021**, *33*, 2003786. [CrossRef] [PubMed]

12. Cherevko, S.; Geiger, S.; Kasian, O.; Kulyk, N.; Grote, J.-P.; Savan, A.; Shrestha, B.R.; Merzlikin, S.; Breitbach, B.; Ludwig, A.; et al. Oxygen and hydrogen evolution reactions on Ru, RuO_2, Ir, and IrO_2 thin film electrodes in acidic and alkaline electrolytes: A comparative study on activity and stability. *Catal. Today* **2016**, *262*, 170–180. [CrossRef]
13. Shi, Q.; Zhu, C.; Du, D.; Lin, Y. Robust noble metal-based electrocatalysts for oxygen evolution reaction. *Chem. Soc. Rev.* **2019**, *48*, 3181–3192. [CrossRef] [PubMed]
14. Jo, S.; Kim, M.-C.; Lee, K.B.; Choi, H.; Zhang, L.; Sohn, J.I. Nonprecious High-Entropy Chalcogenide Glasses-Based Electrocatalysts for Efficient and Stable Acidic Oxygen Evolution Reaction in Proton Exchange Membrane Water Electrolysis. *Adv. Energy Mater.* **2023**, *13*, 2301420. [CrossRef]
15. Kong, S.; Li, A.; Long, J.; Adachi, K.; Hashizume, D.; Jiang, Q.; Fushimi, K.; Ooka, H.; Xiao, J.; Nakamura, R. Acid-stable manganese oxides for proton exchange membrane water electrolysis. *Nat. Catal.* **2024**, *7*, 252–261. [CrossRef]
16. Pu, Z.; Liu, T.; Amiinu, I.S.; Cheng, R.; Wang, P.; Zhang, C.; Ji, P.; Hu, W.; Liu, J.; Mu, S. Transition-Metal Phosphides: Activity Origin, Energy-Related Electrocatalysis Applications, and Synthetic Strategies. *Adv. Funct. Mater.* **2020**, *30*, 2004009. [CrossRef]
17. Yang, L.; Zhou, L.; Wang, M.; Qi, Y.; Guo, D.; Li, H.; Chen, X.; Wang, S. Transition Metal-Based Electrocatalysts for Seawater Oxidation. *Adv. Mater. Interfaces* **2022**, *9*, 2201486. [CrossRef]
18. Guo, T.; Li, L.; Wang, Z. Recent Development and Future Perspectives of Amorphous Transition Metal-Based Electrocatalysts for Oxygen Evolution Reaction. *Adv. Energy Mater.* **2022**, *12*, 2200827. [CrossRef]
19. Han, L.; Dong, S.; Wang, E. Transition-Metal (Co, Ni, and Fe)-Based Electrocatalysts for the Water Oxidation Reaction. *Adv. Mater.* **2016**, *28*, 9266–9291. [CrossRef]
20. Ledendecker, M.; Krick Calderon, S.; Papp, C.; Steinruck, H.P.; Antonietti, M.; Shalom, M. The synthesis of nanostructured Ni5 P4 films and their use as a non-noble bifunctional electrocatalyst for full water splitting. *Angew. Chem. Int. Ed. Engl.* **2015**, *54*, 12361–12365. [CrossRef]
21. Sun, H.; Yan, Z.; Liu, F.; Xu, W.; Cheng, F.; Chen, J. Self-Supported Transition-Metal-Based Electrocatalysts for Hydrogen and Oxygen Evolution. *Adv. Mater.* **2020**, *32*, e1806326. [CrossRef] [PubMed]
22. Kim, B.K.; Kim, M.J.; Kim, J.J. Modulating the active sites of nickel phosphorous by pulse-reverse electrodeposition for improving electrochemical water splitting. *Appl. Catal. B Environ.* **2022**, *308*, 121226. [CrossRef]
23. Liu, X.; Yu, Q.; Qu, X.; Wang, X.; Chi, J.; Wang, L. Manipulating Electron Redistribution in Ni_2P for Enhanced Alkaline Seawater Electrolysis. *Adv. Mater.* **2024**, *36*, e2307395. [CrossRef]
24. Jiang, M.; Zhai, H.; Chen, L.; Mei, L.; Tan, P.; Yang, K.; Pan, J. Unraveling the Synergistic Mechanism of Bi-Functional Nickel–Iron Phosphides Catalysts for Overall Water Splitting. *Adv. Funct. Mater.* **2023**, *33*, 2302621. [CrossRef]
25. Xu, T.; Jiao, D.; Zhang, L.; Zhang, H.; Zheng, L.; Singh, D.J.; Zhao, J.; Zheng, W.; Cui, X. Br-induced P-poor defective nickel phosphide for highly efficient overall water splitting. *Appl. Catal. B Environ.* **2022**, *316*, 121686. [CrossRef]
26. Li, Y.; Wu, Y.; Hao, H.; Yuan, M.; Lv, Z.; Xu, L.; Wei, B. In situ unraveling surface reconstruction of Ni_5P_4@FeP nanosheet array for superior alkaline oxygen evolution reaction. *Appl. Catal. B Environ.* **2022**, *305*, 121033. [CrossRef]
27. Liu, K.; Wang, F.; He, P.; Shifa, T.A.; Wang, Z.; Cheng, Z.; Zhan, X.; He, J. The Role of Active Oxide Species for Electrochemical Water Oxidation on the Surface of 3d-Metal Phosphides. *Adv. Energy Mater.* **2018**, *8*, 1703290. [CrossRef]
28. Menezes, P.W.; Indra, A.; Das, C.; Walter, C.; Göbel, C.; Gutkin, V.; Schmeiβer, D.; Driess, M. Uncovering the Nature of Active Species of Nickel Phosphide Catalysts in High-Performance Electrochemical Overall Water Splitting. *ACS Catal.* **2017**, *7*, 103–109. [CrossRef]
29. Stern, L.-A.; Feng, L.; Song, F.; Hu, X. Ni_2P as a Janus catalyst for water splitting: The oxygen evolution activity of Ni_2P nanoparticles. *Energy Environ. Sci.* **2015**, *8*, 2347–2351. [CrossRef]
30. Kim, B.K.; Kim, S.-K.; Cho, S.K.; Kim, J.J. Enhanced catalytic activity of electrodeposited Ni-Cu-P toward oxygen evolution reaction. *Appl. Catal. B Environ.* **2018**, *237*, 409–415. [CrossRef]
31. Gao, L.; Cui, X.; Sewell, C.D.; Li, J.; Lin, Z. Recent advances in activating surface reconstruction for the high-efficiency oxygen evolution reaction. *Chem. Soc. Rev.* **2021**, *50*, 8428–8469. [CrossRef]
32. Jiang, H.; He, Q.; Li, X.; Su, X.; Zhang, Y.; Chen, S.; Zhang, S.; Zhang, G.; Jiang, J.; Luo, Y.; et al. Tracking Structural Self-Reconstruction and Identifying True Active Sites toward Cobalt Oxychloride Precatalyst of Oxygen Evolution Reaction. *Adv. Mater.* **2019**, *31*, e1805127. [CrossRef] [PubMed]
33. Chen, X.; Cheng, Y.; Wen, Y.; Wang, Y.; Yan, X.; Wei, J.; He, S.; Zhou, J. CoP/Fe-Co_9S_8 for Highly Efficient Overall Water Splitting with Surface Reconstruction and Self-Termination. *Adv. Sci.* **2022**, *9*, e2204742. [CrossRef] [PubMed]
34. Laursen, A.B.; Wexler, R.B.; Whitaker, M.J.; Izett, E.J.; Calvinho, K.U.D.; Hwang, S.; Rucker, R.; Wang, H.; Li, J.; Garfunkel, E.; et al. Climbing the Volcano of Electrocatalytic Activity while Avoiding Catalyst Corrosion: Ni3P, a Hydrogen Evolution Electrocatalyst Stable in Both Acid and Alkali. *ACS Catal.* **2018**, *8*, 4408–4419. [CrossRef]
35. Banerjee, S.; Kakekhani, A.; Wexler, R.B.; Rappe, A.M. Relationship between the Surface Reconstruction of Nickel Phosphides and Their Activity toward the Hydrogen Evolution Reaction. *ACS Catal.* **2023**, *13*, 4611–4621. [CrossRef]

36. Medford, A.J.; Vojvodic, A.; Hummelshøj, J.S.; Voss, J.; Abild-Pedersen, F.; Studt, F.; Bligaard, T.; Nilsson, A.; Nørskov, J.K. From the Sabatier principle to a predictive theory of transition-metal heterogeneous catalysis. *J. Catal.* **2015**, *328*, 36–42. [CrossRef]
37. Wang, B.; Zhang, F. Main Descriptors To Correlate Structures with the Performances of Electrocatalysts. *Angew. Chem. Int. Ed. Engl.* **2022**, *61*, e202111026. [CrossRef]
38. Zhang, W.; Yang, L.; Li, Z.; Nie, G.; Cao, X.; Fang, Z.; Wang, X.; Ramakrishna, S.; Long, Y.; Jiao, L. Regulating Hydrogen/Oxygen Species Adsorption via Built-in Electric Field -Driven Electron Transfer Behavior at the Heterointerface for Efficient Water Splitting. *Angew. Chem. Int. Ed. Engl.* **2024**, *63*, e202400888. [CrossRef]
39. Wexler, R.B.; Martirez, J.M.P.; Rappe, A.M. Active Role of Phosphorus in the Hydrogen Evolving Activity of Nickel Phosphide (0001) Surfaces. *ACS Catal.* **2017**, *7*, 7718–7725. [CrossRef]
40. Wexler, R.B.; Martirez, J.M.P.; Rappe, A.M. Stable Phosphorus-Enriched (0001) Surfaces of Nickel Phosphides. *Chem. Mater.* **2016**, *28*, 5365–5372. [CrossRef]
41. Liyanage, I.A.; Flores, A.V.; Gillan, E.G. Tunable Synthesis of Metal-Rich and Phosphorus-Rich Nickel Phosphides and Their Comparative Evaluation as Hydrogen Evolution Electrocatalysts. *Inorg. Chem.* **2023**, *62*, 4947–4959. [CrossRef]
42. He, J.; Morales-García, Á.; Bludský, O.; Nachtigall, P. The surface stability and equilibrium crystal morphology of Ni2P nanoparticles and nanowires from an ab initio atomistic thermodynamic approach. *CrystEngComm* **2016**, *18*, 3808–3818. [CrossRef]
43. Fei, H.; Dong, J.; Feng, Y.; Allen, C.S.; Wan, C.; Volosskiy, B.; Li, M.; Zhao, Z.; Wang, Y.; Sun, H.; et al. General synthesis and definitive structural identification of MN_4C_4 single-atom catalysts with tunable electrocatalytic activities. *Nat. Catal.* **2018**, *1*, 63–72. [CrossRef]
44. Kuo, D.-Y.; Paik, H.; Kloppenburg, J.; Faeth, B.; Shen, K.M.; Schlom, D.G.; Hautier, G.; Suntivich, J. Measurements of Oxygen Electroadsorption Energies and Oxygen Evolution Reaction on RuO_2(110): A Discussion of the Sabatier Principle and Its Role in Electrocatalysis. *J. Am. Chem. Soc.* **2018**, *140*, 17597–17605. [CrossRef]
45. He, F.; Zheng, Q.; Yang, X.; Wang, L.; Zhao, Z.; Xu, Y.; Hu, L.; Kuang, Y.; Yang, B.; Li, Z.; et al. Spin-State Modulation on Metal-Organic Frameworks for Electrocatalytic Oxygen Evolution. *Adv. Mater.* **2023**, *35*, e2304022. [CrossRef]
46. Bai, X.; Zhang, X.; Sun, Y.; Huang, M.; Fan, J.; Xu, S.; Li, H. Low Ruthenium Content Confined on Boron Carbon Nitride as an Efficient and Stable Electrocatalyst for Acidic Oxygen Evolution Reaction. *Angew. Chem. Int. Ed. Engl.* **2023**, *62*, e202308704. [CrossRef]
47. Chen, Z.; Yang, H.; Mebs, S.; Dau, H.; Driess, M.; Wang, Z.; Kang, Z.; Menezes, P.W. Reviving Oxygen Evolution Electrocatalysis of Bulk La-Ni Intermetallics via Gaseous Hydrogen Engineering. *Adv. Mater.* **2023**, *35*, e2208337. [CrossRef]
48. Ren, X.; Zhai, Y.; Wang, P.; Xu, Z.; Gao, S.; Chen, X.; Gu, Q.; Wang, B.; Li, J.; Liu, S.F. Surface Restructuring of Zeolite-Encapsulated Halide Perovskite to Activate Lattice Oxygen Oxidation for Water Electrolysis. *Adv. Mater.* **2023**, *35*, e2301166. [CrossRef]
49. Li, C.F.; Zhao, J.W.; Xie, L.J.; Wu, J.Q.; Ren, Q.; Wang, Y.; Li, G.R. Surface-Adsorbed Carboxylate Ligands on Layered Double Hydroxides/Metal-Organic Frameworks Promote the Electrocatalytic Oxygen Evolution Reaction. *Angew. Chem. Int. Ed. Engl.* **2021**, *60*, 18129–18137. [CrossRef]
50. Shah, K.; Dai, R.; Mateen, M.; Hassan, Z.; Zhuang, Z.; Liu, C.; Israr, M.; Cheong, W.C.; Hu, B.; Tu, R.; et al. Cobalt Single Atom Incorporated in Ruthenium Oxide Sphere: A Robust Bifunctional Electrocatalyst for HER and OER. *Angew. Chem. Int. Ed. Engl.* **2022**, *61*, e202114951. [CrossRef]
51. Hajiyani, H.; Pentcheva, R. Surface Termination and Composition Control of Activity of the Co_xNi_1–xFe_2O_4(001) Surface for Water Oxidation: Insights from DFT+U Calculations. *ACS Catal.* **2018**, *8*, 11773–11782. [CrossRef]
52. Lv, X.; Wei, W.; Wang, H.; Huang, B.; Dai, Y. Multifunctional electrocatalyst PtM with low Pt loading and high activity towards hydrogen and oxygen electrode reactions: A computational study. *Appl. Catal. B Environ.* **2019**, *255*, 117743. [CrossRef]
53. Xu, Q.; Jiang, H.; Zhang, H.; Hu, Y.; Li, C. Heterogeneous interface engineered atomic configuration on ultrathin $Ni(OH)_2/Ni_3S_2$ nanoforests for efficient water splitting. *Appl. Catal. B Environ.* **2019**, *242*, 60–66. [CrossRef]
54. Zhang, S.L.; Guan, B.Y.; Lu, X.F.; Xi, S.; Du, Y.; Lou, X.W.D. Metal Atom-Doped Co_3O_4 Hierarchical Nanoplates for Electrocatalytic Oxygen Evolution. *Adv. Mater.* **2020**, *32*, e2002235. [CrossRef]
55. Zhao, S.; Wang, Y.; Hao, Y.; Yin, L.; Kuo, C.H.; Chen, H.Y.; Li, L.; Peng, S. Lewis Acid Driving Asymmetric Interfacial Electron Distribution to Stabilize Active Species for Efficient Neutral Water Oxidation. *Adv. Mater.* **2024**, *36*, e2308925. [CrossRef]
56. Hohenberg, P.; Kohn, W. Inhomogeneous Electron Gas. *Phys. Rev.* **1964**, *136*, B864–B871. [CrossRef]
57. Kohn, W.; Sham, L.J. Self-Consistent Equations Including Exchange and Correlation Effects. *Phys. Rev.* **1965**, *140*, A1133–A1138. [CrossRef]
58. Perdew, J.P.; Burke, K.; Ernzerhof, M. Generalized Gradient Approximation Made Simple. *Phys. Rev. Lett.* **1996**, *77*, 3865–3868. [CrossRef] [PubMed]
59. Rappe, A.M.; Rabe, K.M.; Kaxiras, E.; Joannopoulos, J.D. Optimized pseudopotentials. *Phys. Rev. B* **1990**, *41*, 1227–1230. [CrossRef]
60. Ramer, N.J.; Rappe, A.M. Designed nonlocal pseudopotentials for enhanced transferability. *Phys. Rev. B* **1999**, *59*, 12471–12478. [CrossRef]

61. Grimme, S.; Ehrlich, S.; Goerigk, L. Effect of the damping function in dispersion corrected density functional theory. *J. Compu[illegible] Chem.* **2011**, *32*, 1456–1465. [CrossRef] [PubMed]
62. Larsson, E.J.A.K. An X-ray investigation of the Ni-P system and the crystal structures of NiP and NiP_2. *Ark. Kemi* **1965**, *2[illegible]* 335–365.
63. Bengtsson, L. Dipole correction for surface supercell calculations. *Phys. Rev. B* **1999**, *59*, 12301–12304. [CrossRef]
64. Wang, L.; Hao, Y.; Deng, L.; Hu, F.; Zhao, S.; Li, L.; Peng, S. Rapid complete reconfiguration induced actual active species fo[illegible] industrial hydrogen evolution reaction. *Nat. Commun.* **2022**, *13*, 5785. [CrossRef]
65. Zhang, L.; Wang, L.; Wen, Y.; Ni, F.; Zhang, B.; Peng, H. Boosting Neutral Water Oxidation through Surface Oxygen Modulatio[illegible] *Adv. Mater.* **2020**, *32*, e2002297. [CrossRef]
66. Han, M.; Wang, N.; Zhang, B.; Xia, Y.J.; Li, J.; Han, J.R.; Yao, K.L.; Gao, C.C.; He, C.N.; Liu, Y.C.; et al. High-Valent Nicke[illegible] Promoted by Atomically Embedded Copper for Efficient Water Oxidation. *ACS Catal.* **2020**, *10*, 9725–9734. [CrossRef]
67. Nørskov, J.K.; Rossmeisl, J.; Logadottir, A.; Lindqvist, L.; Kitchin, J.R.; Bligaard, T.; Jónsson, H. Origin of the Overpotential fo[illegible] Oxygen Reduction at a Fuel-Cell Cathode. *J. Phys. Chem. B* **2004**, *108*, 17886–17892. [CrossRef]
68. Man, I.C.; Su, H.Y.; Calle-Vallejo, F.; Hansen, H.A.; Martínez, J.I.; Inoglu, N.G.; Kitchin, J.; Jaramillo, T.F.; Norskov, J.K.; Rossmeisl[illegible] J. Universality in Oxygen Evolution Electrocatalysis on Oxide Surfaces. *Chemcatchem* **2011**, *3*, 1159–1165. [CrossRef]
69. Reuter, K.; Scheffler, M. Composition, structure, and stability of RuO_2(110)as a function of oxygen pressure. *Phys. Rev. B* **2001**, *65*[illegible] 035406. [CrossRef]

catalysts

MDPI

Article

Tailoring β-Bi_2O_3 Nanoparticles via Mg Doping for Superior Photocatalytic Activity and Hydrogen Evolution

Ibrahim M. Sharaf [1], Mohamed S. I. Koubisy [1], Fatemah H. Alkallas [2], Amira Ben Gouider Trabelsi [2] and Abdelaziz Mohamed Aboraia [1,*]

1 Physics Department, Faculty of Science, Al-Azhar University, Assiut 71524, Egypt
2 Department of Physics, College of Science, Princess Nourah Bint Abdulrahman University, P.O. Box 84428, Riyadh 11671, Saudi Arabia
* Correspondence: a.m.aboraia@gmail.com or a.m.aboraia@azhar.edu.eg

Abstract: Bismuth oxide (β-Bi_2O_3) is a promising visible-light-driven photocatalyst due to its narrow direct bandgap, but its practical application is hindered by rapid electron–hole recombination and limited surface active sites. This study demonstrates a sol-gel synthesis approach to tailor β-Bi_2O_3 nanoparticles through magnesium (Mg) doping, achieving remarkable enhancements in the photocatalytic degradation of organic pollutants and hydrogen evolution. The structural analysis through XRD, SEM, and EDX confirmed Mg-doping concentrations of 0.025 to 0.1 M led to crystallite size reduction from 79 nm to 13 nm, while the UV–Vis bandgap measurement showed it decreased from 3.8 eV to 3.08–3.3 eV. The photodegradation efficiency increased through Mg doping at a 0.1 M concentration, with the highest rate constant value of 0.0217 min^{-1}. The doping process led to VB potential reduction between 3.37 V (pristine) and 2.78–2.91 V across the doped samples when referenced to SCE. The photocatalytic performance of $Mg_{0.075}Bi_{1.925}O_3$ improved with its 3.2 V VB potential because the photoelectric band arrangement enhanced both light absorption and charge separation. The combination of modifications through Mg doping yielded an enhanced photocatalytic performance, which proves that magnesium doping is a pivotal approach to modifying β-Bi_2O_3 suitable for environmentally and energy-related applications.

Keywords: nanoparticles; Mg-doped Bi_2O_3; methylene blue; kinetics; hydrogen evaluation; photodegradation

1. Introduction

Extensive research into advanced wastewater treatment approaches has grown due to two parallel factors: rising global water cleanliness needs, together with environmental pollution concerns [1,2]. Semiconductor-based photocatalysis represents a leading wastewater degradation technology because it achieves efficient organic pollutant elimination through sustainable solar-powered solutions [3–5]. Bismuth oxide (Bi_2O_3) stands as a popular semiconductor material that scientists have widely examined because of its appropriate bandgap energy, durable photostability, and strong visible light sensitivity [6]. The practical usage of pure Bi_2O_3 faces two main challenges since it recycles electron–hole pairs at high speed and allows minimal sunlight absorption [7,8]. Different researchers have examined the photocatalytic performance of Bi_2O_3 by implementing three main strategies that included doping with other elements, along with oxygen vacancy development and nanostructure design [9–11].

The process of incorporating metal ions functions as a successful technique for altering both the electronic characteristics and optical properties of semiconductor materials [12,13].

The alkaline earth metal magnesium represents a viable dopant option since it introduces both defect states and enables specific crystal phase stabilization [14,15]. The introduction of Mg into the material structure creates oxygen vacancies that act as primary facilitators for increased charge-carrier separation and improved photocatalytic performance [16,17]. When Mg is incorporated into the Bi_2O_3 lattice structure, it affects the bandgap properties, which enables more efficient visible light absorption, ideal for solar-powered applications [7].

β-Bi_2O_3 demonstrates excellent potential for photocatalysis because of its narrow bandgap (~2.3–2.8 eV) and strong absorption of visible light [18,19]. The β-phase has a short-lived existence since it transforms to alternate phases when heat exposure increases, thus restricting its potential practical applications [20,21]. Bismuth oxide (Bi_2O_3) functions as an effective photocatalyst under visible light irradiation, owing to its direct bandgap of 2.8 eV [7]. Research indicates that its photocatalytic efficiency can be significantly enhanced through metal ion doping [22,23]. For instance, Hameed et al. [24] demonstrated improved degradation of methyl orange, methylene blue, and phenol using Zn-modified Bi_2O_3. Similarly, Anandan et al. [25] explored the photocatalytic behavior of Au–Bi_2O_3 nanorods. Additionally, Xie et al. [26] reported that Bi_2O_3 doped with transition metals such as vanadate (VV), lead (PbII), silver (AgI), and cobalt (CoII) exhibited superior photocatalytic activity compared to undoped Bi_2O_3 under visible light irradiation [7]. Mg doping stabilizes the β-phase structure of Bi_2O_3 by making it suitable for photocatalytic utilization. Mg doping leads to oxygen-vacancy formation, which creates new locations for pollutant interaction as well as enhancing general photocatalytic capabilities [27,28].

Methylene blue (MB) serves as a widespread organic dye used in textile and printing operations, which substantially pollutes water [29,30]. The durability of methylene blue in water bodies leads to significant environmental dangers, together with health risks for humans [31]. The removal methods for methylene blue utilizing adsorption and biological treatment effectively fail to provide suitable results while creating additional pollutants [32]. Photocatalysis serves as a sustainable treatment method because it transforms organic dyes into harmless substances such as CO_2 and H_2O through light-powered reactions [29,33]. The development of efficient, cost-effective sunlight-responsive photocatalysts exists as an essential obstacle.

This work describes how new Mg-doped β-Bi_2O_3 nanoparticles with oxygen vacancies were synthesized through a simple sol-gel combustion approach. Systematic investigation of the Mg incorporation in β-Bi_2O_3 lattices used different analytical tools, including X-ray diffraction and scanning electron microscopy. Herein, we studied the bandgap modifications together with the optical properties of nanoparticles through UV–Vis absorption spectroscopy. An evaluation of Mg-doped β-Bi_2O_3 nanoparticles took place through the use of MB solutions under sunlight illumination for photocatalytic testing. The laboratory results show that β-Bi_2O_3's photocatalytic behavior improves because Mg-doping brings together three key effects, which boost charge-carrier separation and oxygen vacancy formation and extend the light absorption range. This study presents novel insights into using alkaline earth metals for improving photocatalytic material performance while demonstrating Mg-doped β-Bi_2O_3 nanoparticles as an effective, sustainable photocatalyst for environmental cleanup. Furthermore, this work aims to create advanced enhanced photocatalytic materials which combine Bi_2O_3 methods with Mg-doping benefits to address real applications in solar energy use and wastewater treatment.

2. Results and Discussion

2.1. Structural Study of Bi_2O_3 and Mg-Doped Bi_2O_3

The XRD patterns of Bi_2O_3 doped with magnesium exist at various molar ratios as shown in Figure 1. The prepared samples demonstrate high crystallinity because of their distinct reflectivity peak pattern. The number of diffraction peaks decreased as the magnesium content in the solution grew. The XRD patterns from both undoped and doped samples matched the tetragonal β-phase Bi_2O_3 structure, which belongs to the space group P-421c (JCPDS Card No. 27-0050), while displaying lattice parameters of a = 7.7425 Å and c = 5.6313 Å. Analysis through X-ray diffraction confirmed complete phase purity for the synthesized samples, since the Mg ion peaks as well as other impurity peaks were absent. The tetragonal Bi_2O_3 crystal structure showed its main characteristic peaks at angular positions which matched with the reflections of the diffraction planes (−1 1 1), (0 2 0), (−1 0 2), (0 0 2), (−1 1 2), (−1 2 1), (−2 0 2), (−2 1 2), (−1 1 3), (0 4 1), (−1 0 4), (−2 4 1), and (−2 2 4). The results of this study match those from past studies that discuss how Bi_2O_3 allotropes form differently based on synthesis temperature ranges [7,34]. The different radii of the ions between bismuth and magnesium indicate that magnesium inserts itself partly into places where bismuth ions exist in the Bi_2O_3 crystal framework. XRD analysis demonstrates that very little Mg substitution of Bi occurs, although the crystal compositions show minimal differences.

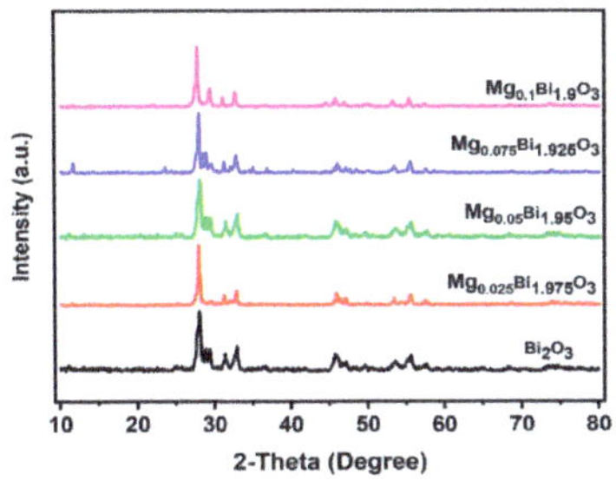

Figure 1. XRD patterns of $Bi_{2-x}Mg_xO_3$ (x = 0, 0.025, 0.05, 0.075, and 0.1).

The average crystallite size and microstrain ε were calculated using the W.H. equation (Equation (1)) based on the most intense XRD peak [35].

$$\beta \cos\theta = \frac{k\lambda}{D} + 4\varepsilon\, sin\theta \quad (1)$$

Here, k = 0.9 represents the numerical shape factor, a constant value; D denotes the crystallite size; λ is the wavelength of the incident radiation; β is the full width at half maximum (FWHM) in radians; and θ is the Bragg angle measured in radians.

After knowing the crystallite size, the dislocation density δ of the material was estimated as given in Equation (2).

$$\delta = \frac{1}{D^2} \quad (2)$$

Table 1 indicates the materials crystallite size, the lattice strain, and the dislocation densities. The introduction of magnesium and different crystallite sizes explained the variable lattice constants and dislocation densities. Table 1 shows that the crystallite size and microstrain reduced with the increasing incorporation of Mg in Bi_2O_3 until 5% of Mg doping; however, the increase up to 7.5% Mg doping led to an increase in the size and strain, as demonstrated in Figure 2. Moreover, the increase up to 10% of Mg doping allowed a decrease in size to around 30 nm may be due to lattice distortion and strain relaxation caused by the smaller ionic radius of Mg^{2+}; the inhibition of crystal growth due to the introduction of defects and boundaries; the formation of oxygen vacancies and defects that promote smaller crystallite formation; the stabilization of the β-phase leading to a more uniform and less strained crystal structure; and the enhanced nucleation rate, favoring the formation of smaller crystallites. These combined effects result in a reduction in both crystallite size and microstrain as the Mg-doping concentration increases.

Figure 3a–c exhibits the FESEM images of how Mg-doped Bi_2O_3 particles become smaller in size while their morphology changes with increasing Mg material concentration. The transformation of crystal growth and nucleation properties and the development of lattice defects follow the effects of Mg doping in Bi_2O_3. The higher Mg concentrations enable small particle dimensions as well as superior surface properties which benefit photocatalyst systems that require large surface areas and many defects. The FSEM-EDX mapping technique of the $Bi_{1.9}Mg_{0.1}O_3$ material demonstrates the distribution of Bi, Mg, and O with smaller particles showing rough surface characteristics due to the presence of Mg atoms, as shown in Figure 4a–d. The uniform distribution of Mg between the Bi_2O_3 lattice crystals verifies the successful incorporation of Mg while no impurity phases indicate the synthesized material remains pure. The observations line up with the anticipated Mg doping outcomes that cause $Bi_{1.9}Mg_{0.1}O_3$ to become a material suitable for photocatalyst applications through its smaller crystallites, increased defects, and better surface properties.

Table 1. The calculated crystalline size, microstrain, dislocation density, and energy gap via change in concentrations of Mg-doped Bi_2O_3 nanoparticles.

Samples	2θ, °	D-Crystalline Size (nm)	Microstrain (ε)	Dislocation Density × 10^{-4} (nm^{-2})	Bandgap Energy (eV)
Bi_2O_3	27.943 29.076 31.393 32.772 32.772 45.852 47.015 53.599 55.408 57.453 74.290	79.7	74.23	1.8145	3.8
$Mg_{0.025}Bi_{1.975}O_3$	27.943 29.076 31.393 32.772 32.772 45.852 47.015 53.599 55.408 57.453 74.290	18.1	36.741	7.407	3.08
$Mg_{0.05}Bi_{1.95}O_3$	27.943 29.076 31.393 32.772 32.772 45.852 47.015 53.599 55.408 57.453 74.290	43.9	35.307	8.0217	3.13

Table 1. *Cont.*

Samples	2θ, °	D-Crystalline Size (nm)	Microstrain (ε)	Dislocation Density × 10^{-4} (nm^{-2})	Bandgap Energy (eV)
$Mg_{0.075}Bi_{1.925}O_3$	27.943 29.076 31.393 32.772 32.772 45.852 47.015 53.599 55.408 57.453 74.290	27.2	45.096	4.91714	3.3
$Mg_{0.1}Bi_{1.9}O_3$	27.943 29.076 31.393 32.772 32.772 45.852 47.015 53.599 55.408 57.453 74.290	12.3	29.542	0.00115	3.14

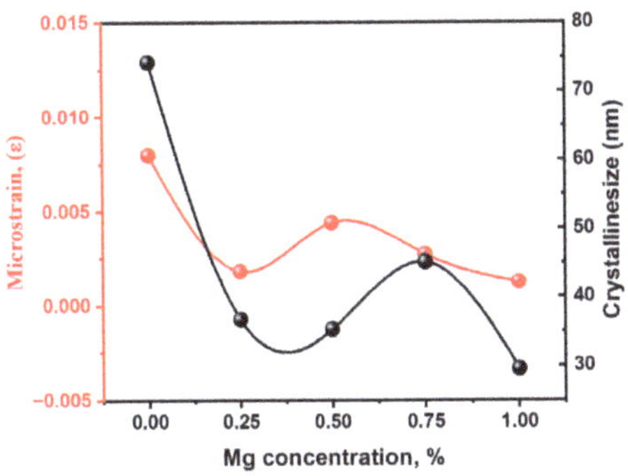

Figure 2. Variation in crystallite size and microstrain of $Bi_{2-x}Mg_xO_3$ (x = 0, 0.025, 0.05, 0.075, and 0.1).

Figure 3. FSEM photos of (**a**) Bi_2O_3, (**b**) $Bi_{1.975}Mg_{0.025}O_3$, and (**c**) $Bi_{1.9}Mg_{0.1}O_3$.

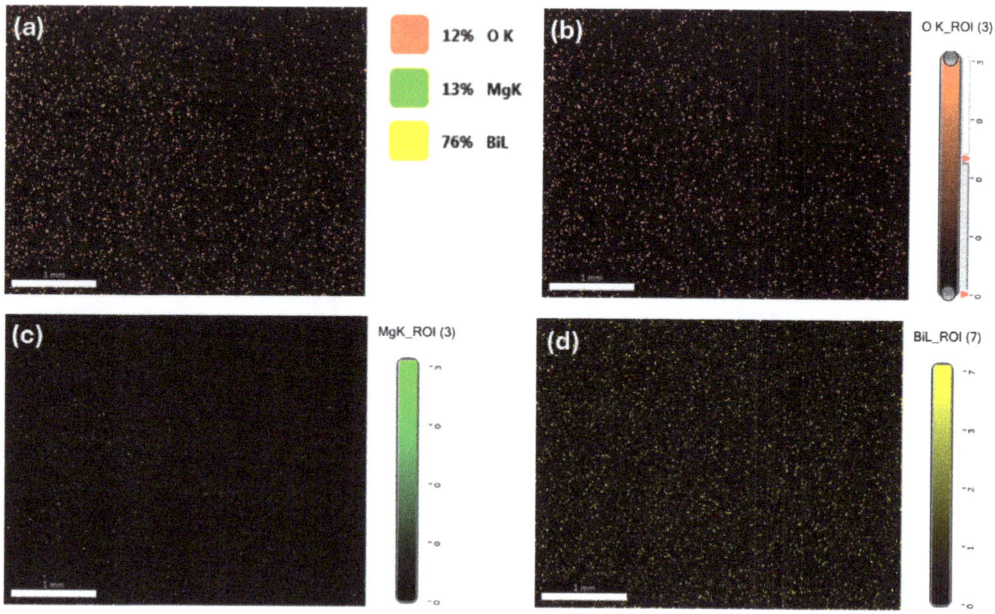

Figure 4. FSM-EDX mapping of $Bi_{1.9}Mg_{0.1}O_3$.

2.2. UV–Visible Spectroscopy

Catalyst material responses toward UV or visible light illumination and their light absorption properties, which activate catalyst particles, need thorough examination. The maximum absorbance measurement of Mg-doped Bi_2O_3 nanoparticles used UV–Vis spectroscopy through exposure to wavelengths from 200 to 800 nm. Figure 5a shows that pure Bi_2O_3 absorbs visible light with a maximum absorption edge at 300 nm, but Mg-doped Bi_2O_3 nanoparticles absorb light at a longer wavelength of 425 nm. When Mg is used to modify Bi_2O_3, the material shows two distinct peaks located at 290 and 350 nm. Although pure Bi_2O_3 shows the same level of absorption as Mg-doped Bi_2O_3, the dopant material demonstrates stronger absorption across the visible light spectrum, suggesting its potential for use in photocatalytic systems. The improved light absorption capability of this material enhances visible light usage for the generation of additional charge carriers from photon absorption. Mg incorporation and the production of oxygen vacancies in the bandgap enhance Mg-doped Bi_2O_3 nanoparticles by broadening their wavelength absorption capabilities and raising their absorption intensity. Oxygen vacancies in Mg-doped Bi_2O_3 introduce shallow-trap levels inside the bandgap that trap excited electrons according to Singh [36].

Determining the optical bandgap energy (Eg) is essential to understanding which portion of the solar spectrum can be absorbed by a semiconductor. The optical bandgap (Eg) was calculated by analyzing the fundamental absorption associated with the excitation of electrons from the valence band to the conduction band. The bandgap (Eg) can be derived from a plot of $(\alpha h\nu)^2$ versus photon energy ($h\nu$) using the well-known Tauc relation [37]:

$$(\alpha h\nu)^2 = A\left(h\nu - E_g\right) \tag{3}$$

where α is the absorption coefficient, A is a constant, Eg is the energy gap, and ν is the light frequency, respectively.

The energy gaps were determined by drawing $(\alpha h\nu)^2$ versus $h\nu$ plots, followed by extrapolation of the linear parts to the x-axis. The series of bandgaps extending from pristine Bi_2O_3 to magnesium-doped Bi_2O_3 (0.025 to 0.1 M) provided various values of 3.8 eV, 3.08 eV, 3.13 eV, 3.3 eV, and 3.14 eV. The results showed a decrease in the energy gap size of magnesium-doped samples when compared to pure Bi_2O_3.

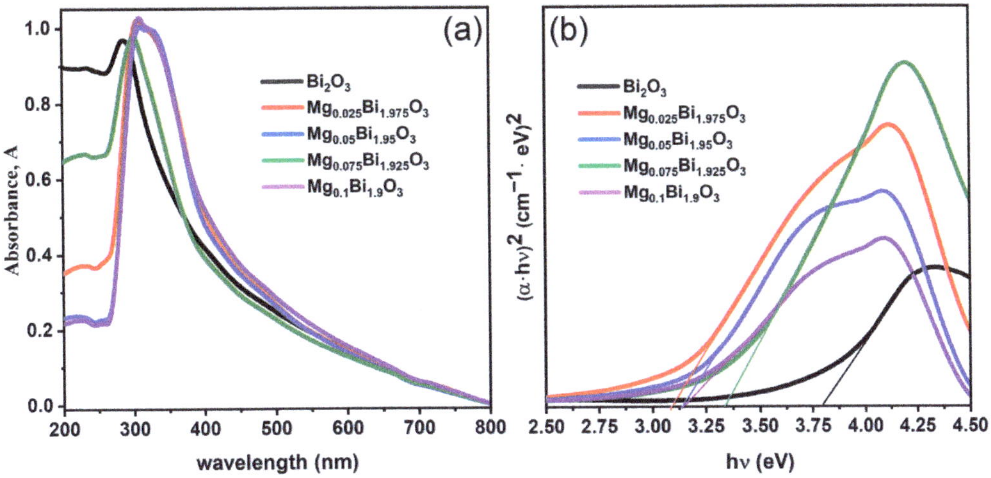

Figure 5. (**a**) The ultraviolet–visible (UV–Vis) absorption spectrum of $Bi_{2-x}Mg_xO_3$; (**b**) Tauc's plot of pure bismuth oxide and various concentrations of Mg-doped Bi_2O_3.

During these experiments, the bandgap values generated corresponded to the findings documented by Wang et al. [38]. Umebayashi et al. [39] demonstrated that dopant introduction creates occupied electronic states within the valence band or bandgap, along with a dependency on alkaline earth metal oxidation states and the composition of their cationic sublattices. This research added Mg^{2+} ions into the material structure, which generated impurity states in the forbidden zone that modified both donor energy levels above the valence band and acceptor energy levels below the conduction band. Bi_2O_3 experienced a decrease in its bandgap to 3.08 eV because of this occurrence. The photocatalytic properties of Bi_2O_3 become stronger after magnesium doping because the resulting bandgap reduction improves its visible light sensitivity.

2.3. Photocatalytic Activity

The spectrum demonstrates that MB exhibits a powerful initial absorption at 665 nm. Undoped Bi_2O_3 exhibits a slow peak intensity reduction over time because of its restricted photoactivity. Powdered Bi_2O_3 exhibits a slow reaction speed because its wide bandgap prevents it from efficiently absorbing visible light and creating electron–hole pairs. Pure Bi_2O_3 demonstrates weak photocatalytic activity since MB decomposition reaches a small extent after 120 min of exposure to radiation, as shown in Figure 6a. The MB absorption peak starting at 665 nm shows similar strength to that of the undoped material. The photocatalytic activity increases in the sample because the peak reduction happens faster than in pure Bi_2O_3. The performance enhancement comes from Mg introducing defects that shrink the bandgap and enable better visible light absorption and charge-carrier separation. Low Mg doping leads to a major reduction in MB over 120 min, thus demonstrating its positive effects. The UV–Vis absorption spectra confirm that dye-degradation efficiency increases with rising Mg concentration to reach its highest point, except for the 0.75 Mg doping, as shown in Figure 6c. The photocatalytic improvements come from narrow bandgap structures and improved visible light absorption ability while producing efficient charge-carrier separation. Tests have established the value of Mg-doped Bi_2O_3 as a highly potent photocatalytic agent under visible light conditions for environmental cleansing applications, as demonstrated in Figure 6d.

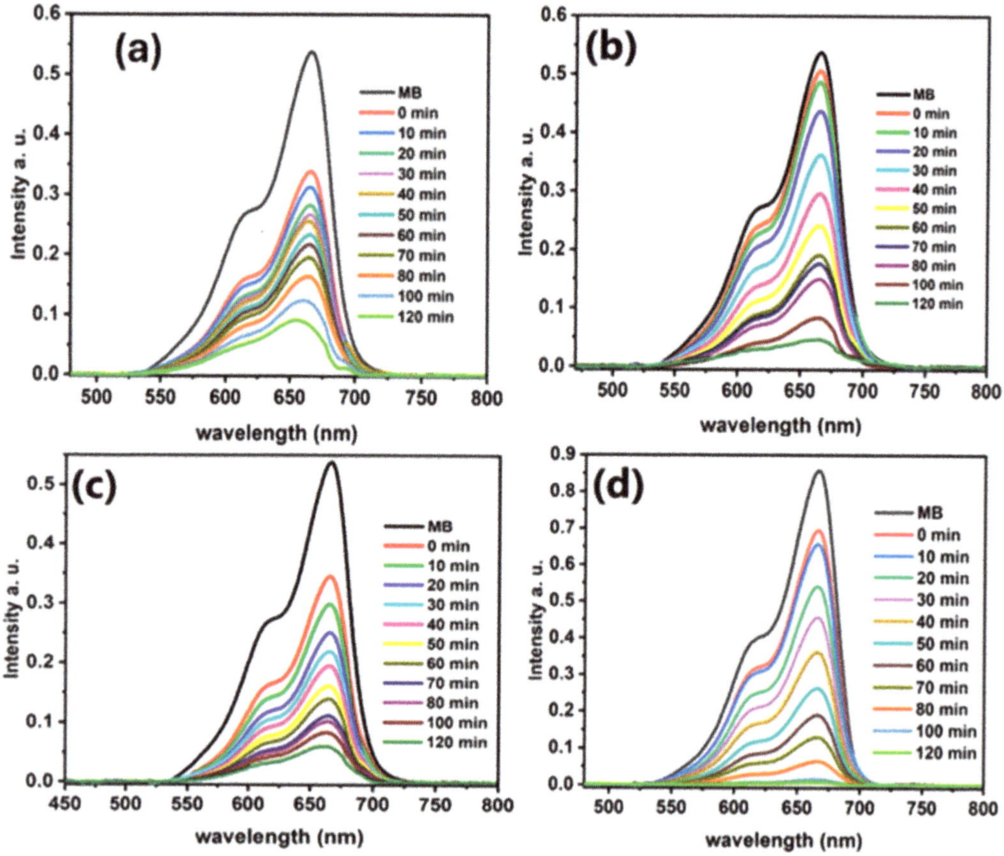

Figure 6. UV–Vis absorption spectra for the degradation of MB dye after irradiation for two hours of the prepared (**a**) 0, (**b**) 0.025, (**c**) 0.05, and (**d**) 0.1 Mg-doped Bi_2O_3.

The efficacy of both undoped and Mg-doped Bi_2O_3 photocatalysts increases proportionally with the amount of Mg-doping element; the best performance belongs to the sample with the highest Mg content. The time-dependent MB concentration changes provided an assessment of the efficiency, as shown in Figure 7. When Mg is incorporated into Bi_2O_3, it forms new electronic states that enable additional interband sites. The photocatalyst surface maintains effective charge-carrier separation because electrons become trapped inside sites that create new electronic states along the bandgap. The presence of oxygen vacancies in Mg-doped Bi_2O_3 impedes electron (e^-) and hole (h^+) recombination because they enhance the separation of photogenerated charge carriers. The photocatalytic efficiency of 0.5% Mg-doped Bi_2O_3 proves to be lower than other samples which indicates excessive electron–hole recombination takes place. Greater amounts of Mg in the doping process enable enhanced photoproduction of electrons and holes which boosts photocatalytic reaction efficiency.

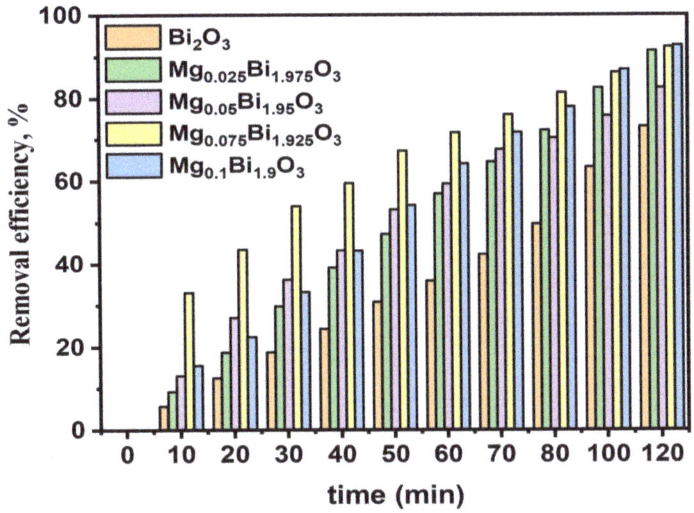

Figure 7. Contrast the degradation effectiveness of a catalyst at various concentrations of magnesium (Mg) doping with bismuth oxide (Bi_2O_3) under simulated sunlight.

2.4. Kinetics Study for the Degradation of MB

The photodegradation of dyes follows pseudo-first-order kinetics and is expressed in the following equation:

$$Ln\left(\frac{C_o}{C_t}\right) = K_1 \cdot t \tag{4}$$

C_o and C_t stand for the initial and remaining concentrations of MB at the particular illumination time before determining K_1, which represents the first-order rate constant. Research data on dye degradation underwent linear least squares analysis to obtain the primary parameters k and R^2. The experimental data fitted precisely against the model through the presented method, which produced accurate parameter values from the analysis in Figure 8. The model shows excellent effectiveness for describing how the dye reacts to light exposure through its strong correlation with experimental data. Figure 8 shows that Mg/Bi_2O_3 NPs displayed maximum photocatalytic activity by producing the highest K value compared to single Bi_2O_3 and other doping samples. The rate constants k exist as listed values in Table 2. The rate constants determined through pseudo-first-order analysis corresponded with experimental photodegradation output at x = 0, 0.025, 0.05, 0.075, and 0.1, resulting in values of 0.0105, 0.02, 0.0146, 0.0182, and 0.0217, respectively. The photocatalytic efficiency reaches its peak when Mg doping achieves its optimal concentration level according to the increasing rate constant data. The fitting accuracy of experimental data by the pseudo-first-order model appears through its R^2 coefficient of determination. The pseudo-first-order model prediction strength indicated by R^2 was higher in ($\ln(C_o/C_t)$ vs. t) than the corresponding value from the pseudo-second-order model (1/Ct vs. t). The data variances explainable by the model reflect its strong representative power for degradation behavior through the high R^2 value.

The research demonstrates that Bi_2O_3 nanoparticles with added Mg perform outstandingly when used to break down methylene blue and other organic pollutants through photocatalytic processes. The material demonstrates exceptional properties because it possesses a narrow bandgap as well as improved charge-carrier separation and structural robustness, which allows its use in prolonged wastewater applications. Congruent research indicates that Mg doping establishes oxygen vacancies for charge preservation, which form small energy levels between the valence band and conduction band states. Photogenerated electrons follow a new excitation path by moving from the VB to oxygen vacancy states while leaving behind holes in the VB. Additionally, the position of the CB in Mg-Bi_2O_3 facilitates the formation of superoxide radicals (O^{2-}) from molecular oxygen in the solution. The photoexcited holes from the catalyst function as highly oxidizing agents, which form hydroxyl radicals (OH) alongside oxygen near the surface. Reactive radicals originating from the catalyst break organic methylene blue molecules in their chemical bonds to produce fragmentary products.

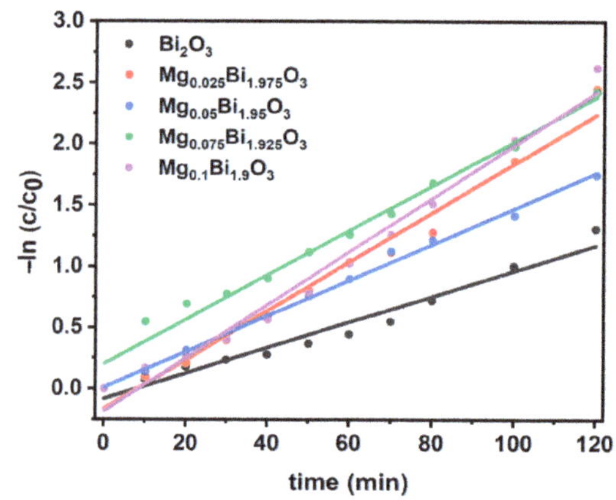

Figure 8. An examination of methylene blue (MB) degradation kinetics using $Bi_{2-x}Mg_xO_3$ at various doping levels (x = 0, 0.025, 0.05, 0.075, and 0.1) employing pseudo-first-order kinetic models.

Table 2. Parameters related to pseudo-first-order kinetic models for the photodegradation of methylene blue using $Bi_{2-x}Mg_xO_3$ with varying dopant concentrations (x = 0, 0.025, 0.05, 0.075, and 0.1) as the photocatalyst.

Samples	Removal Efficiency, %	k	R^2
Bi_2O_3	73	0.0105	0.96
$Mg_{0.025}Bi_{1.975}O_3$	91	0.02	0.979
$Mg_{0.05}Bi_{1.95}O_3$	82	0.0146	0.995
$Mg_{0.075}Bi_{1.925}O_3$	92	0.0182	0.98
$Mg_{0.1}Bi_{1.9}O_3$	93	0.0217	0.979

The evaluation of different catalyst efficiencies for separating charge carriers used photocurrent measurements. The chronoamperometric test of photoelectrochemical experiments yielded results that are shown in Figure 9a. The serial on-off changes operated for five seconds revealed a direct and stable current transformation in each switch. The sample $Mg_{0.075}Bi_{1.925}O_3$ showed a remarkably lower photocurrent than each of its composition counterparts. Another parameter measured from the photoelectrochemical experiments was photocurrent density represented by 7, 9, 13, 40, and 12 $\mu A/cm^2$ for Bi_2O_3, $Mg_{0.025}Bi_{1.975}O_3$, $Mg_{0.05}Bi_{1.95}O_3$, and $Mg_{0.075}Bi_{1.925}O_3$, respectively. Photocurrent density analysis showed that $Mg_{0.025}Bi_{1.975}O_3$ achieved the most extensive photocurrent value of 40 $\mu A/cm^2$, surpassing every other sample tested.

The findings confirm that heterostructure formation, the Bi_2O_3 electron sink nature, and Mg incorporation enhance charge-carrier separation efficiency as well as electron-transfer rates. Mott–Schottky measurements confirmed the band structure of Bi_2O_3 along with its Mg-doped variants $Mg_{0.025}Bi_{1.975}O_3$, $Mg_{0.05}Bi_{1.95}O_3$, $Mg_{0.075}Bi_{1.925}O_3$, and $Mg_{0.1}Bi_{1.9}O_3$, thus validating the previous results. The researchers determined freely mobile charge-carrier concentrations that were equivalent to normal-condition donor densities, after considering Mg-doping effects on phase transitions. Measurement procedures took place under dark conditions at a 500 Hz measurement frequency. Figure 9b demonstrates that pure Bi_2O_3 functions as an n-type semiconductor, given its positive slope in the Mott–Schottky analysis. EIS spectra of Bi_2O_3 and fabricated $Mg_{0.025}Bi_{1.975}O_3$, $Mg_{0.05}Bi_{1.95}O_3$, $Mg_{0.075}Bi_{1.925}O_3$, and $Mg_{0.1}Bi_{1.9}O_3$ samples are presented in Figure 9c. Posterior to Bi_2O_3, the radius of the EIS arc diminishes sequentially through the following materials: $Mg_{0.075}Bi_{1.925}O_3$, then $Mg_{0.05}Bi_{1.95}O_3$, then $Mg_{0.1}Bi_{1.9}O_3$, culminating in $Mg_{0.025}Bi_{1.975}O_3$. The semicircle diameter in the Nyquist plot acts as an inverse indicator for material charge transport efficiency. The smaller proportion of the EIS arc demonstrates useful characteristics, including better photogenerated charge separation and accelerated interfacial charge routes, and slower electron–hole (e^-/h^+) pair recombination. The optimal electrical conductivity emerges from the $Mg_{0.075}Bi_{1.925}O_3$ compound as revealed in Figure 9c through its minimal charge-transfer resistance values. The optimized Mg doping creates conditions that lead to the reduced carrier recombination of photogenerated carriers. Among the tested samples, $Mg_{0.075}Bi_{1.925}O_3$ stands out as the most effective photocatalyst because it shows superior performance in morphological analysis, together with structural and optoelectronic analysis. The authors selected $Mg_{0.075}Bi_{1.925}O_3$ for detailed examination because this composition demonstrated the best performance.

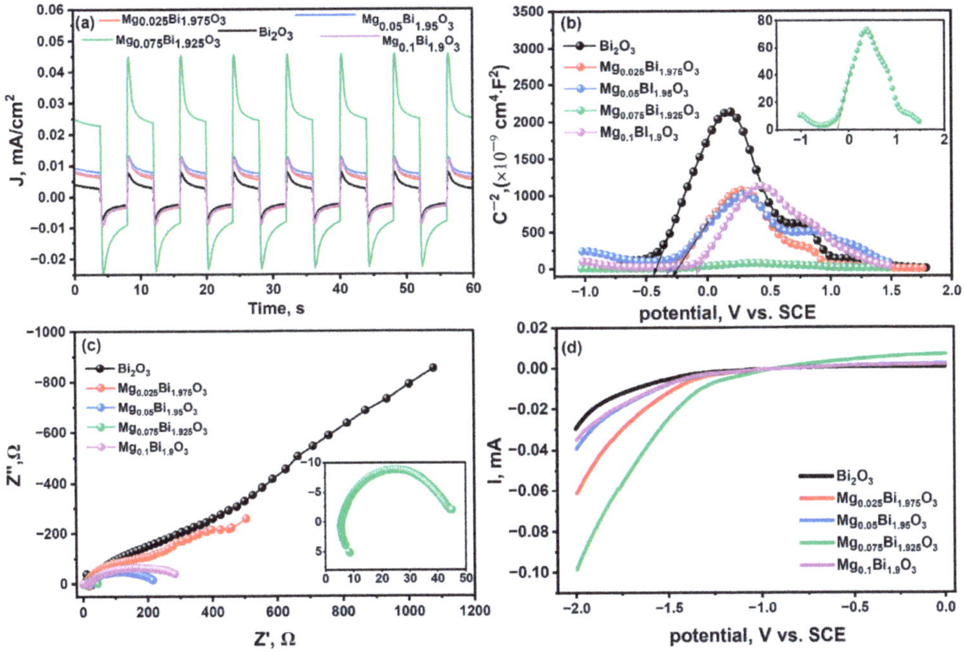

Figure 9. (**a**) Transient photocurrent responses, (**b**) Mott–Schottky, (**c**) EIS, and (**d**) LSV polarization curves of Bi_2O_3, $Mg_{0.025}Bi_{1.975}O_3$, $Mg_{0.05}Bi_{1.95}O_3$, $Mg_{0.075}Bi_{1.925}O_3$, and $Mg_{0.1}Bi_{1.9}O_3$ nanocomposite samples.

The efficiency of photocatalysts used for hydrogen production depends heavily on the value of their hydrogen evolution reaction potential (HER). A photocatalytic system for H_2 generation shows better prospects when the overpotential remains low. A linear sweep voltammetry analysis illustrates that $Mg_{0.075}Bi_{1.925}O_3$ possesses the most favorable hydrogen evolution overpotential of −1.34 V vs. SCE among Bi_2O_3, $Mg_{0.025}Bi_{1.975}O_3$, $Mg_{0.05}Bi_{1.95}O_3$, $Mg_{0.075}Bi_{1.925}O_3$, and $Mg_{0.1}Bi_{1.9}O_3$. $Mg_{0.075}Bi_{1.925}O_3$ demonstrates the most suitable conditions for photocatalytic H_2 production because of its optimal HER overpotential value of −1.34 V vs. SCE when compared to virgin Bi_2O_3 with a greater overpotential of −1.87 V vs. SCE. The obtained bandgap structures for all samples stemmed from Tauc's plot analysis and are displayed as conduction band (CB) values in Figure 10. The valence band (VB) potential readings for the samples Bi_2O_3, $Mg_{0.025}Bi_{1.975}O_3$, $Mg_{0.05}Bi_{1.95}O_3$, $Mg_{0.075}Bi_{1.925}O_3$, and $Mg_{0.1}Bi_{1.9}O_3$ were 3.37 V, 2.8 V, 2.78 V, 3.2 V, and 2.91 V (vs. SCE), respectively, with CB values. The optimized band arrangement of $Mg_{0.075}Bi_{1.925}O_3$ helps to enhance the efficiency of charge separation and light trapping which leads to a better performance in breaking down organic pollutants through photocatalysis.

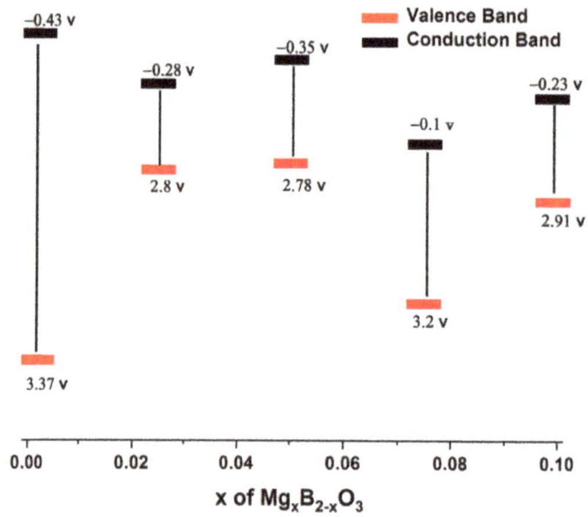

Figure 10. Band diagram derived by assuming the flat band potential of the films as the potential of the conduction band (CB). The valence band (VB) position was estimated by adding the measured bandgap energy.

3. Materials and Methods

3.1. Synthesis of Mg-Doped Bismuth Oxide NPs

Bi_2O_3 nanoparticles were prepared by the solution combustion method using urea as fuel. A cylindrical crucible took the stoichiometric quantities of a mixture containing $Bi(NO_3)_3 \cdot 5H_2O$ (99.999% Loba Co., Mumbai, India) and fuel urea [CH_4N_2O] (99.999% Loba Co. India). The mixture in the crucible was stirred well to obtain homogeneity for 10 min at 400 rpm, and a clear, transparent solution was formed. This crucible was placed in a muffle furnace and heated to 600 °C for two hours. The resulting product was cooled to room temperature overnight and collected. A yellow powder was obtained, crushed using a pestle and mortar, and characterized. To prepare the samples for doping with different concentrations of Mg, the same procedure was followed for undoped Bi_2O_3 as for doping with varying ratios of molar $Mg(NO_3)_3 \cdot 9H_2O$ (99.999% Loba Co., India). The molar ratios of Mg/Bi were 0%, 0.25%, 0.5%, 0.75%, and 1%, respectively. Finally, a pale yellow powder with an increase in Mg^{2+} content was easily ground into a powder.

3.2. Characterization

The crystalline structure and morphology of a series of $Bi_{2-x}Mg_xO_3$ (x = 0, 0.025, 0.05, 0.075, and 0.1) samples were considered by employing XRD, SEM, EDX, and X-ray diffraction (XRD) analysis was performed using a LANScientific instrument (Suzhou, China) with a wavelength (λ = 0.15418 nm) with diffraction peaks recorded between 10° and 80° (2θ), and a step of 0.06 degree/s was exploited to investigate the crystal structure. The morphology, particle size, and elemental concentration were investigated using a scanning electron microscope (SEM, a JEOL JSM 5910, Tokyo, Japan) with an accelerating voltage of 20 kV. Energy-dispersive X-ray (EDX) spectroscopy (JSM-5910, Tokyo, Japan) was used to study the manufactured material's compositional features. The bandgap energy was measured with a JASCO 670 UV–visible spectrophotometer (Tsukuba-shi, Japan). The acquired materials were examined for their photocatalytic efficacy using organic dye pollutants, specifically methylene blue (MB). The electrochemistry measurements were carried out by Corretest Cs 305 (Wuhan, China). To perform the photocatalysis experiment, 20 milligrams of photocatalyst material was combined with a 500-watt xenon lamp.

3.3. Photocatalytic Performance of $Bi_{2-x}Mg_xO_3$

The photocatalytic assessment of synthesized Bi_2O_3 particles occurred through the investigation of their ability to break down methylene blue (MB) in water. The UV–Vis spectrophotometer (JASCO 670) determined that the MB absorbance peak existed within the visible spectrum range from 400 to 800 nm while analyzing the absorbance spectrum. The reaction monitoring required the usage of 665 nm as the wavelength where the maximum absorbance was detected. The experiments used visible light illumination, which included a Pyrex-glass-made immersion well photochemical reactor as the light source. A cylindrical reactor contained the light source and had a water-cooling jacket for experimental temperature management. Each test used 0.02 g of photocatalyst, either from Bi_2O_3 or Mg-doped Bi_2O_3, with 50 mL of MB solution at a 10 ppm concentration, which was stirred at 600 rpm. The solution underwent 30 min sonication under dark conditions before irradiation for the achievement of adsorption–desorption equilibrium. The chemical degradation of the dye solution was measured through absorbance detection at 665 nm. The photocatalysis process indicated its dye-removal efficiency through the implementation of this specific calculation [40,41]:

$$Removal\ \% = \frac{(C_o - C_t)}{C_o} \times 100 \tag{5}$$

After resolving the complex process of photocatalytic degradation, the initial (C_0) and final (C_t) concentrations of the pollutant are accurately measured. The catalyst is separated via centrifugation and then re-dispersed in the dye solution for further recycling evaluations.

4. Conclusions

The sol-gel synthesized Mg-doped β-Bi_2O_3 nanoparticles brought about significant photocatalytic enhancements because of their optimized structural properties along with their adjusted electronic features. The addition of Mg doping at concentrations between 0.025 and 0.1 M resulted in crystallite sizes decreasing from 79 nm (pristine Bi_2O_3) down to 13 nm, together with a decrease in bandgap from 3.8 eV to 3.08–3.3 eV. The modified bandgap and optimized valence band potential levels between 2.78 and 2.91 V vs. SCE improved the separation of charges and absorption of sunlight. The sample with $Mg_{0.075}Bi_{1.925}O_3$ composition displayed the optimal band alignment structure that enabled excellent light-trapping abilities and superior charge-transfer capabilities which resulted in a high organic pollutant photodegradation rate constant of 0.0217 min^{-1} and superior hydrogen evolution performance. Research indicates that Mg doping represents an essential method to design β-Bi_2O_3 materials for the continued development of sustainable environmental cleanup systems and renewable energy solutions. The researchers should study scalability and long-term stability for practical implementation in future investigations.

Author Contributions: Conceptualization, A.B.G.T. and A.M.A.; Methodology, I.M.S., M.S.I.K., F.H.A. and A.M.A.; Validation, M.S.I.K. and A.B.G.T.; Formal analysis, I.M.S. and A.B.G.T.; Investigation, I.M.S. and F.H.A.; Resources, M.S.I.K.; Data curation, I.M.S., M.S.I.K., A.B.G.T. and A.M.A.; Writing—original draft, F.H.A., A.B.G.T. and A.M.A.; Visualization, A.M.A.; Supervision, A.M.A. All authors have read and agreed to the published version of the manuscript.

Funding: This research was funded by Princess Nourah bint Abdulrahman University grant number PNURSP2025R38.

Data Availability Statement: The data presented in this study are available on request from the corresponding author.

Acknowledgments: The authors express their gratitude to Princess Nourah bint Abdulrahman University Researchers Supporting Project Number PNURSP2025R38, Princess Nourah bint Abdulrahman University, Riyadh, Saudi Arabia.

Conflicts of Interest: The authors declare no conflict of interest.

References

1. Aboraia, A.M.; Al-omoush, M.; Solayman, M.; Saad, H.M.H.; Khabiri, G.; Saad, M.; Alsulaim, G.M.; Soldatov, A.V.; Ismail, Y.A.M.; Gomaa, H. A heterostructural MoS2QDs@UiO-66 nanocomposite for the highly efficient photocatalytic degradation of methylene blue under visible light and simulated sunlight. *RSC Adv.* **2023**, *13*, 34598–34609. [CrossRef] [PubMed]
2. Aboraia, A.M.; Almohammedi, A.; Alraddadi, S.; Taha, S.A.; Saad, M.; Sharaf, I.; Ismail, Y.A.M. Comparison study between as-synthesized ZnO and ZnO derived from ZiF-8 metalorganic framework in removing methylene blue. *Mod. Phys. Lett. B* **2024**, *38*, 2450221. [CrossRef]
3. Ahmed, M.; Mavukkandy, M.O.; Giwa, A.; Elektorowicz, M.; Katsou, E.; Khelifi, O.; Naddeo, V.; Hasan, S.W. Recent developments in hazardous pollutants removal from wastewater and water reuse within a circular economy. *NPJ Clean Water* **2022**, *5*, 12. [CrossRef]
4. Ahmed, M.A.; Mohamed, A.A. Recent progress in semiconductor/graphene photocatalysts: Synthesis, photocatalytic applications, and challenges. *RSC Adv.* **2023**, *13*, 421–439. [CrossRef]
5. Ajmal, A.; Majeed, I.; Malik, R.N.; Iqbal, M.; Nadeem, M.A.; Hussain, I.; Yousaf, S.; Mustafa, G.; Zafar, M.I.; Nadeem, M.A. Photocatalytic degradation of textile dyes on Cu_2O-CuO/TiO_2 anatase powders. *J. Environ. Chem. Eng.* **2016**, *4*, 2138–2146. [CrossRef]
6. Tudor, M.; Borlan, R.; Maniu, D.; Astilean, S.; de la Chapelle, M.L.; Focsan, M. Plasmon-enhanced photocatalysis: New horizons in carbon dioxide reduction technologies. *Sci. Total Environ.* **2024**, *932*, 172792. [CrossRef]
7. Malathy, P.; Vignesh, K.; Rajarajan, M.; Suganthi, A. Enhanced photocatalytic performance of transition metal doped Bi_2O_3 nanoparticles under visible light irradiation. *Ceram. Int.* **2014**, *40*, 101–107. [CrossRef]
8. Zhang, L.; Niu, J.; Li, D.; Gao, D.; Shi, J. Preparation and Photocatalytic Activity of Ag Modified Ti-Doped-Bi_2O_3 Photocatalyst. *Adv. Condens. Matter Phys.* **2014**, *2014*, 749354.
9. Mane, V.; Dake, D.; Raskar, N.; Sonpir, R.; Stathatos, E.; Dole, B. A review on Bi_2O_3 nanomaterial for photocatalytic and antibacterial applications. *Chem. Phys. Impact* **2024**, *8*, 100517. [CrossRef]
10. Dou, W.; Hu, X.; Kong, L.; Peng, X. Photo-induced dissolution of Bi_2O_3 during photocatalysis reactions: Mechanisms and inhibition method. *J. Hazard. Mater.* **2021**, *412*, 125267. [CrossRef]
11. He, W.; Sun, Y.; Jiang, G.; Huang, H.; Zhang, X.; Dong, F. Activation of amorphous Bi_2WO_6 with synchronous Bi metal and Bi_2O_3 coupling: Photocatalysis mechanism and reaction pathway. *Appl. Catal. B Environ.* **2018**, *232*, 340–347. [CrossRef]
12. Li, J.; Zhang, J.Z. Optical properties and applications of hybrid semiconductor nanomaterials. *Coord. Chem. Rev.* **2009**, *253*, 3015–3041. [CrossRef]
13. Jiang, R.; Li, B.; Fang, C.; Wang, J. Metal/semiconductor hybrid nanostructures for plasmon-enhanced applications. *Adv. Mater.* **2014**, *26*, 5274–5309. [CrossRef] [PubMed]
14. Zheng, B.; Fan, J.; Chen, B.; Qin, X.; Wang, J.; Wang, F.; Deng, R.; Liu, X. Rare-earth doping in nanostructured inorganic materials. *Chem. Rev.* **2022**, *122*, 5519–5603. [CrossRef] [PubMed]
15. Phung, N.; Félix, R.; Meggiolaro, D.; Al-Ashouri, A.; Sousa e Silva, G.; Hartmann, C.; Hidalgo, J.; Köbler, H.; Mosconi, E.; Lai, B. The doping mechanism of halide perovskite unveiled by alkaline earth metals. *J. Am. Chem. Soc.* **2020**, *142*, 2364–2374. [CrossRef]
16. Duan, X.; Liu, N. Magnesium for dynamic nanoplasmonics. *Acc. Chem. Res.* **2019**, *52*, 1979–1989. [CrossRef]
17. Chen, Y.; Zhou, W.; Ding, D.; Liu, M.; Ciucci, F.; Tade, M.; Shao, Z. Advances in cathode materials for solid oxide fuel cells: Complex oxides without alkaline earth metal elements. *Adv. Energy Mater.* **2015**, *5*, 1500537. [CrossRef]

18. Zhong, F.-G.; Huang, X.-Y.; Barreto, R.; Chen, S.-H. Synthetization and photocatalytic performance of high-aspect-ratio β-Bi_2O_3 nanowires via the combination of vacuum die-casting and controllable oxidation processes in liquid phase. *Ceram. Int.* **2025**, *12*, 15802–15808. [CrossRef]
19. Li, J.; Fu, M.; Wu, Z.; Chu, P.; Wei, C.; Zhang, Y.; Huang, M.; Yang, Y.; He, D.; Wang, Y. Versatile fabrication of spherical inverse opals from diverse materials including Bi_2O_2/g-C_2N_4 and MXene/H-TiO_2 using hierarchical microspheres as templates. *Ceram. Int.* **2024**, *5*, 6253–6263.
20. Chavan, K.M.; Mane, V.A.; Raskar, N.D.; Dake, D.V.; Sonpir, R.B.; Dole, B.N. Nanostructured MnO_2-Based Zn-Doped Bi_2O_3 Nanocomposite for Improved Antimicrobial and Photocatalytic Applications. *ChemistrySelect* **2024**, *9*, e202403806. [CrossRef]
21. Fedyaeva, O.N.; Grebennikov, A.P.; Vostrikov, A.A. The role of water in the oxidation of lead and bismuth by high-density water-oxygen fluid. *J. Supercrit. Fluids* **2025**, *218*, 106509. [CrossRef]
22. Li, L.; Yan, B. $BiVO_4$/Bi_2O_3 submicrometer sphere composite: Microstructure and photocatalytic activity under visible-light irradiation. *J. Alloys Compd.* **2009**, *476*, 624–628. [CrossRef]
23. Jing, L.; Wang, J.; Qu, Y.; Luan, Y. Effects of surface-modification with Bi_2O_3 on the thermal stability and photoinduced charge property of nanocrystalline anatase TiO_2 and its enhanced photocatalytic activity. *Appl. Surf. Sci.* **2009**, *256*, 657–663. [CrossRef]
24. Hameed, A.; Gombac, V.; Montini, T.; Felisari, L.; Fornasiero, P. Photocatalytic activity of zinc modified Bi_2O_3. *Chem. Phys. Lett.* **2009**, *483*, 254–261. [CrossRef]
25. Anandan, S.; Lee, G.-J.; Chen, P.-K.; Fan, C.; Wu, J.J. Removal of Orange II Dye in Water by Visible Light Assisted Photocatalytic Ozonation Using Bi_2O_3 and Au/Bi_2O_3 Nanorods. *Ind. Eng. Chem. Res.* **2010**, *49*, 9729–9737. [CrossRef]
26. Xie, J.; Lü, X.; Chen, M.; Zhao, G.; Song, Y.; Lu, S. The synthesis, characterization and photocatalytic activity of V(V), Pb(II), Ag(I) and Co(II)-doped Bi_2O_3. *Dye. Pigment.* **2008**, *77*, 43–47. [CrossRef]
27. Hou, Y.; An, X.; An, W.; Wang, H.; Hu, J.; Cui, W. Oxygen vacancy-enhanced magnesium-doped ferrous molybdate photocatalytic-peroxymonosulfate synergistic degradation of pollutants. *Appl. Surf. Sci.* **2024**, *645*, 158908. [CrossRef]
28. Hao, L.; Huang, H.; Zhang, Y.; Ma, T. Oxygen vacant semiconductor photocatalysts. *Adv. Funct. Mater.* **2021**, *31*, 2100919. [CrossRef]
29. Abou-Gamra, Z.M.; Ahmed, M.A. Synthesis of mesoporous TiO_2-curcumin nanoparticles for photocatalytic degradation of methylene blue dye. *J. Photochem. Photobiol. B Biol.* **2016**, *160*, 134–141. [CrossRef]
30. Zhang, X.; Sathiyaseelan, A.; Zhang, L.; Lu, Y.; Jin, T.; Wang, M.-H. Zirconium and cerium dioxide fabricated activated carbon-based nanocomposites for enhanced adsorption and photocatalytic removal of methylene blue and tetracycline hydrochloride. *Environ. Res.* **2024**, *261*, 119720. [CrossRef]
31. Trandafilović, L.V.; Jovanović, D.J.; Zhang, X.; Ptasińska, S.; Dramićanin, M.D. Enhanced photocatalytic degradation of methylene blue and methyl orange by ZnO:Eu nanoparticles. *Appl. Catal. B Environ.* **2017**, *203*, 740–752. [CrossRef]
32. Jing, H.-P.; Wang, C.-C.; Zhang, Y.-W.; Wang, P.; Li, R. Photocatalytic degradation of methylene blue in ZIF-8. *RSC Adv.* **2014**, *4*, 54454–54462. [CrossRef]
33. Chanu, L.A.; Singh, W.J.; Singh, K.J.; Devi, K.N. Effect of operational parameters on the photocatalytic degradation of Methylene blue dye solution using manganese doped ZnO nanoparticles. *Results Phys.* **2019**, *12*, 1230–1237. [CrossRef]
34. Luo, X.; Zhu, G.; Peng, J.; Wei, X.; Hojamberdiev, M.; Jin, L.; Liu, P. Enhanced photocatalytic activity of Gd-doped porous β-Bi_2O_3 photocatalysts under visible light irradiation. *Appl. Surf. Sci.* **2015**, *351*, 260–269. [CrossRef]
35. Yahia, I.S.; Shapaan, M.; Ismail, Y.A.M.; Aboraia, A.M.; Shaaban, E.R. Thickness dependence of structural and optical properties of cadmium iodide thin films. *J. Alloys Compd.* **2015**, *636*, 317–322. [CrossRef]
36. Singh, S.; Sahoo, R.K.; Shinde, N.M.; Yun, J.M.; Mane, R.S.; Kim, K.H. Synthesis of Bi_2O_3-MnO_2 nanocomposite electrode for wide-potential window high performance supercapacitor. *Energies* **2019**, *12*, 3320. [CrossRef]
37. Aboraia, A.M.; Yahia, I.S.; Saad, M.; Alsulaim, G.; Alnahdi, K.M.; Alsharif, S.A.; Elewa, N.N.; Ismail, Y.A.M.; Mostafa khalefa, M.; Madkhali, N.; et al. Exploration of the structural rGO thin films and their optical characteristics for optoelectronic device applications. *J. Opt.* **2024**. [CrossRef]
38. Wang, R.; Chu, C.; Wang, L.; Liu, Q. Investigation of the photocatalytic performance of Mg-doped modified BiOCl. *Inorg. Chem. Commun.* **2023**, *157*, 111286. [CrossRef]
39. Umebayashi, T.; Yamaki, T.; Itoh, H.; Asai, K. Analysis of electronic structures of 3d transition metal-doped TiO_2 based on band calculations. *J. Phys. Chem. Solids* **2002**, *63*, 1909–1920. [CrossRef]

40. Aboraia, A.M.; Sharaf, I.M.; Saad, M.; Alsulaim, G.; Alnahdi, K.M.; Elewa, N.N.; Khalefa, M.M.; Ismail, Y.A.M. Exploring the photocatalytic characteristics of samarium doping on the structural and morphological attributes and photocatalytic performance of zirconium dioxide. *Mod. Phys. Lett. B* **2024**, *39*, 2450503. [CrossRef]
41. Alharshan, G.A.; Aboraia, A.M.; Uosif, M.A.M.; Sharaf, I.M.; Shaaban, E.R.; Saad, M.; Almohiy, H.; Elsenety, M.M. Optical band gap tuning, DFT understandings, and photocatalysis performance of ZnO nanoparticle-doped Fe compounds. *Materials* **2023**, *16*, 2676. [CrossRef] [PubMed]

catalysts

MDPI

Article

Synergistic MoS_2–Gold Nanohybrids for Sustainable Hydrogen Production

Shrouq H. Aleithan [1,*], Shroq S. Laradhi [2], Kawther Al-Amer [2] and Hany M. Abd El-Lateef [2,*]

1 Department of Physics, College of Science, King Faisal University, P.O. Box 400, Al-Ahsa 31982, Saudi Arabia
2 Department of Chemistry, College of Science, King Faisal University, P.O. Box 400, Al-Ahsa 31982, Saudi Arabia
* Correspondence: salaithan@kfu.edu.sa (S.H.A.); hmahmed@kfu.edu.sa (H.M.A.E.-L.)

Abstract: Extensive research has been conducted on the catalytic properties of molybdenum disulfide (MoS_2) materials in the context of the hydrogen evolution reaction (HER). This study focuses on exploring hybrid MoS_2/Au structures as a catalyst for HER, utilizing linear sweep voltammetry as the experimental methodology. Firstly, 2D-MoS_2 flakes were synthesized by the chemical vapor deposition (CVD) approach and directly added to gold nanoparticles during or after their preparation process. The prepared nanocomposites were characterized by Fourier transform infrared spectroscopy (FTIR), X-ray diffraction (XRD), Raman spectroscopy, and scanning electron microscopy with energy-dispersive X-ray analysis (SEM/EDX). The HER performance was tested for the two resulting samples to show that the preparation of gold nanoparticles with the coexistence of CVD-MoS_2 flakes produces a superior electrocatalytic performance of the sample in a neutral medium. Notably, the onset potential was measured as −0.152 V (versus reversible hydrogen electrode (RHE)) with an exchange current density (j_0) of 0.22 mA/cm^2. Chronoamperometric data show that all composites retained initial current densities for 15 hours, confirming stable, efficient HER performance post-decay.

Keywords: MoS_2 flaks; Au nanoparticles; electrocatalysis; hydrogen evolution reaction

1. Introduction

As the challenges of energy shortages, environmental degradation, and pollution persist, there has recently been increased interest in exploring clean and sustainable energy sources, such as hydrogen. Electrochemical water splitting technology stands out as a highly efficient and promising approach for environmentally friendly hydrogen production. This process comprises a cathodic hydrogen evolution reaction (HER) and an anodic oxygen evolution reaction (OER) [1]. The electrocatalyst plays a crucial role in facilitating these reactions and minimizing power consumption [2–8]. Conventional electrocatalysts predominantly rely on nanomaterials derived from rare precious metals. For the hydrogen evolution reaction (HER), platinum (Pt), ruthenium (Ru), rhodium (Rh), and palladium (Pd) are widely used due to their high catalytic activity. In contrast, materials such as ruthenium dioxide (RuO_2) and iridium dioxide (IrO_2) are primarily employed for the oxygen evolution reaction (OER) owing to their excellent oxidative stability and efficiency. However, their constrained availability and elevated costs have impeded broad-scale industrial implementation. Consequently, there has been growing interest in transition metal dichalcogenides (TMDs), particularly molybdenum disulfide (MoS_2), valued for its cost-effectiveness and efficacy as an electrocatalyst for the HER. MoS_2 possesses a hexagonal

 https://doi.org/10.3390/catal15060550

tri-layered structure characterized by van der Waals interactions between individual tri-layers, employing versatile applications in energy storage, semiconductors, biomedicine, and electrocatalysis [2,9,10].

In response to the catalytic activity observed at the edge sites of MoS_2, recent research has focused on either enhancing active edge site generation or improving the reactivity of the inert basal planes. Strategies such as chemical methods, including doping, and the manipulation of structural properties, such as crystallinity and nano-structuring, have been explored to expose more active edge sites. However, achieving these desired structures often requires using energy-intensive techniques, including atomic layer deposition, chemical vapor deposition, and hydrothermal methods. An alternative strategy involves the integration of MoS_2 with other materials to form hybrids with enhanced electrical conductivity [2,11–13].

Metal nanoparticles, such as platinum (Pt), palladium (Pd), nickel (Ni) and gold (Au), have been successfully integrated into MoS_2 nanosheets to enhance their catalytic activity, particularly in the hydrogen evolution reaction (HER) [14]. Pt and Pd incorporation have proven effective in improving MoS_2 activity, while MoS_2 decorated with Au nanoparticles has demonstrated exceptional efficacy for the HER. Gold nanoparticles have demonstrated unique catalytic properties, especially at the nanoscale, including high selectivity, excellent stability, and resistance to toxicity, particularly in reactions involving oxygen-containing functional groups. These properties make gold catalysts preferable in some hybrid material systems where long-term stability and low activation energy are critical, even with high raw material costs [15–20]. The notable effectiveness of Au-decorated MoS_2 is attributed to the robust Au-S interaction and the outstanding stability of the gold nanoparticles. This targeted decoration results in an improvement in the charge transport properties of MoS_2, consequently leading to an enhancement in the catalytic efficiency for the hydrogen evolution reaction (HER) [15–20].

In this study, molybdenum disulfide (MoS_2) nanosheets were first synthesized using the chemical vapor deposition (CVD) technique, followed by different methods to incorporate gold nanoparticles (Au NPs). The primary objective of this research was to explore the influence of Au nanoparticle intercalation on the hydrogen evolution reaction (HER) activity of MoS_2. To characterize the sample, a comprehensive set of analytical techniques was employed, including FTIR, XRD, Raman spectroscopy, and SEM/EDX. These analyses aimed to provide a detailed understanding of the structural and compositional aspects of the synthesized nanocomposites. Subsequently, the HER activity of each specimen was evaluated, revealing a significant improvement in the performance of the hydrogen evolution reaction. This improvement was attributed to the presence of Au nanoparticles, which contributed to the increased surface area and improved electrical conductivity of MoS_2.

2. Results and Discussion

2.1. FTIR Spectra

Figure 1a shows absorption peaks in a range from 470 to 1400 cm^{-1}, which are related to MoS_2. The peaks near 1600 cm^{-1} and ~ 3000 cm^{-1} are associated with hydroxyl groups' stretching and bending vibrations. The peak at 1400 cm^{-1} corresponds to the stretching vibration of S–Mo–S bonds. The Mo–S stretching vibration mode is between 490 and 620 cm^{-1} [21–23].

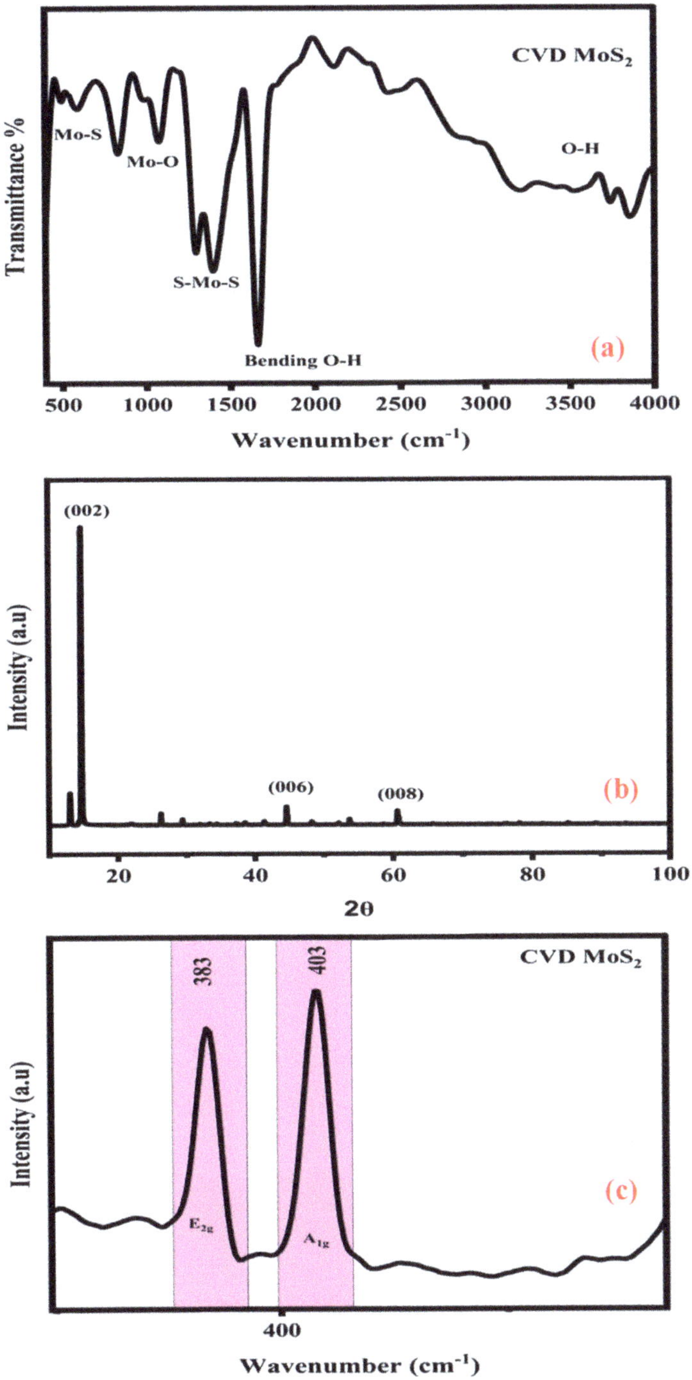

Figure 1. (**a**) The FTIR spectrum of the CVD-MoS_2 composite. (**b**) The XRD pattern of the CVD-MoS_2 composite. (**c**) The Raman spectrum of the CVD-MoS_2 composite.

2.2. XRD Spectra

The X-ray diffraction (XRD) analysis of MoS_2 powder synthesized via chemical vapor deposition (CVD) confirms its crystalline structure. The XRD patterns, measured within the 2θ range of 10°–60°, exhibit diffraction peaks characteristic of 2H-MoS_2, as identified by ICCD card numbers 37-1492. Peaks appear at 2θ values of 14.6°, 44.3°, and 60°, corresponding to the (002), (006) and (008) crystallographic planes, respectively [24–31].

The sharp and narrow XRD peaks suggest that the samples possess high crystallinity (Figure 1b) [32–34].

2.3. Raman Spectra

Raman spectroscopy serves as a reliable and non-destructive method for analyzing lattice vibrations and phonon modes in materials. Figure 1c presents the Raman spectra of the as-synthesized CVD-MoS_2, showing two prominent peaks at approximately 383 cm^{-1} and 403 cm^{-1}, corresponding to the E_{2g} (in-plane) and A_{1g} (out-of-plane) vibrational modes, respectively. The 20 cm^{-1} separation between these peaks confirms the monolayer nature of MoS_2 flakes. The E_{2g} mode arises from opposite vibrations of Mo and S atoms in the plane, while the A_{1g} mode is attributed to sulfur atoms vibrating out-of-plane. The sharp and intense E_{2g} and A_{1g} peaks reflect high crystallinity and the formation of the 2H phase [28,35].

2.4. Dynamic Light Scattering (DLS)

Dynamic light scattering (DLS) was employed to assess the size distribution of gold nanoparticles (AuNPs) synthesized via the Turkevich method, revealing an average hydrodynamic diameter of approximately 26 nm (Figure 2).

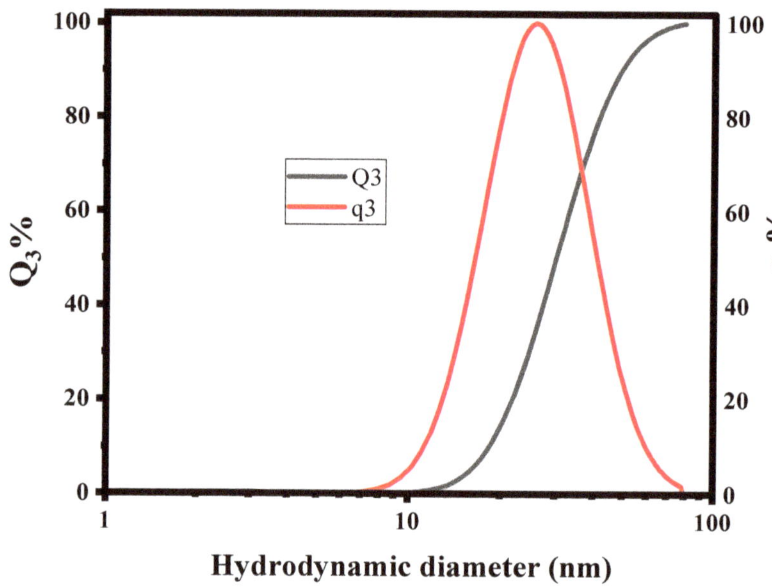

Figure 2. Dynamic light scattering of Au nanoparticles: (Q3) Cumulative Volume Distribution, and (q3) Volume Density Distribution.

2.5. SEM and EDS

The morphology of the sample was examined using scanning electron microscopy (SEM) at various magnifications, as shown in Figure 3. The SEM images (Figure 3a,b) illustrate the existence of expansive flakes characterized by irregular morphologies. These flakes display a micro-scale lateral dimension spanning from 1 to 100 μm, elucidating the superior crystalline structure and well-defined XRD peaks observed in this sample (Figure 1b).

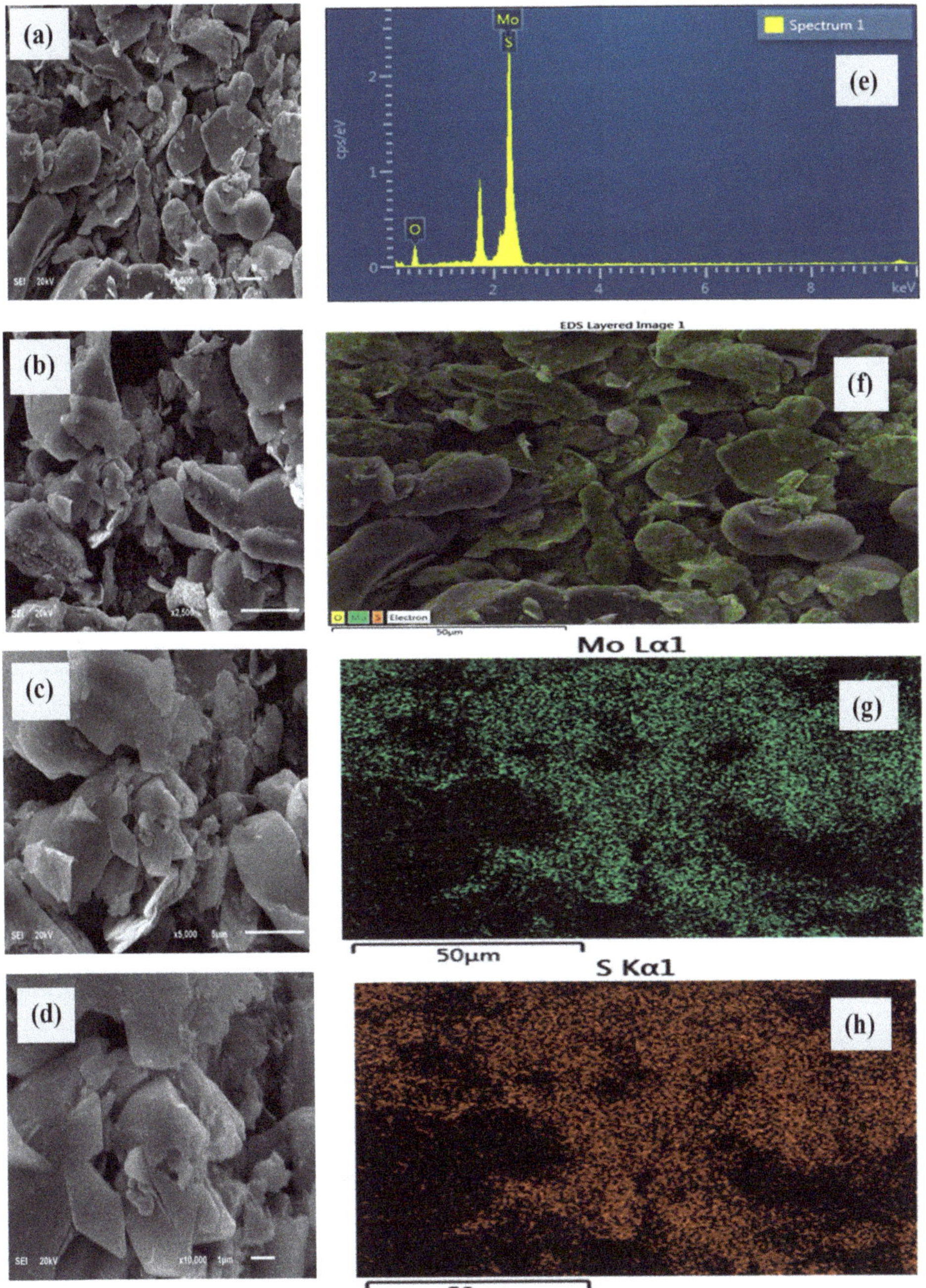

Figure 3. SEM images of the CVD-MoS_2 sample were captured at various magnifications, denoted as (**a–d**). (**e**) The EDS spectrum of CVD-MoS_2. (**f**) The EDS map of CVD-MoS_2. EDS maps of Mo and S elements in samples (**g**) and (**h**), respectively.

EDX data were utilized to determine the elemental composition of the samples (Figure 3c, Table 1). The analysis revealed prominent peaks indicating the presence of Mo and S. Nevertheless, the utilization of the EDX spectrum was not viable for accurately determining the precise composition ratio due to the overlapping peaks of S (K = 2.31 keV) and Mo (L = 2.29 keV) [36].

Table 1. Atomic composition of MoS_2 flakes.

Element	At. No.	Mass [%]	Atom [%]
Oxygen	8	28.88	2.24
Sulfur	8	22.35	1.59
Molybdenum	42	48.76	2.63
Sum		100.00	

2.6. HER Activity of CVD-MoS_2/Au in Neutral Medium

In this study, the hydrogen evolution reaction (HER) performances of CVD-MoS_2, CVD-MoS_2 + Au, and CVD-MoS_2 + prepared Au electrodes in a 1M KCl electrolyte are assessed, and the results are shown in Figure 4. In a neutral medium, the CVD-MoS_2 electrode exhibits an onset potential of 0.215 V (Figure 4a) at a current density of 1 mA/cm^2. The maximum current density achieved is 4.3 mA/cm^2 at an overpotential of 0.372 V, with an exchange current density (j_0) of 0.149 mA/cm^2 at 0.0 V. The addition of prepared gold to CVD-MoS_2 results in an increase in the onset potential to 0.289 V, with a cathodic current of 5.2 mA/cm^2 and a j_0 of 0.186 mA/cm^2. The CVD-MoS_2 + Au electrode shows the best HER performance, with an onset potential of 0.152 V (Figure 4a) at 1 mA/cm^2, a j_0 of 0.22 mA/cm^2, and a maximum current density of 8.4 mA/cm^2 at an overpotential of 0.372 V.

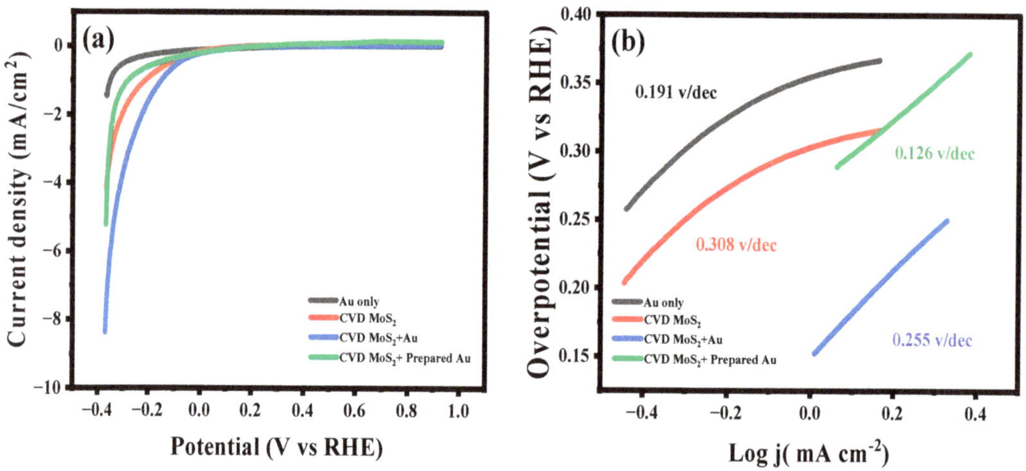

Figure 4. HER activity in neutral medium: (**a**) polarization curve; (**b**) Tafel plot of the Au, CVD-MoS_2, CVD-MoS_2 + Au, and CVD-MoS_2 + prepared Au composites in KCl.

The Tafel slopes, presented in Figure 4b, are 0.308 V/dec for CVD-MoS_2, 0.126 V/dec for CVD-MoS_2 + prepared Au, and 0.255 V/dec for CVD-MoS_2 + Au. The CVD-MoS_2 + prepared Au catalyst demonstrates the fastest HER kinetics, as evidenced by the lowest Tafel slope of 0.126 V/dec (Table 2).

Table 2. The electrochemical parameters were calculated to evaluate the HER performances of all the prepared MoS_2 catalysts in KCl.

Catalyst	Onset Potential (V)	Cathodic Current * (mA/ cm ^{2})	Tafel (sV/dec)	Exchange Current mA cm^{-2} (j_0)
Au	0.357	1.4	0.191	0.111
CVD-MoS_2	0.215	4.3	0.308	0.149
CVD-MoS_2 + prepared Au	0.289	5.2	0.126	0.186
CVD-MoS_2 + Au	0.152	8.4	0.255	0.22
		* at 0.372 V		

As shown in Figure 5a, the polarization curve for CVD-MoS_2 (red curve) reveals an onset potential—defined here as the potential at which a noticeable catalytic current starts to appear, specifically when the current density exceeds 1.0 mA/cm^2—of 0.311 V, accompanied by a steep rise in current density as the overpotential increases. The current density increases with rising overpotential, reaching 2.60 mA/cm^2 within the measured range. The exchange current density (j_0), corresponding to the current density at 0.0 V, is 0.11 mA/cm^2, indicating moderate intrinsic HER activity. Upon the incorporation of Au nanoparticles, the onset potential is reduced to 0.304 V, and the cathodic current improves significantly, reaching 7.48 mA/cm^2 for CVD-MoS_2 + prepared Au and 7.27 mA/cm^2 for CVD-MoS_2 + Au, while enhancing the j_0 value to around 0.199 mA/cm^2 (green curve).

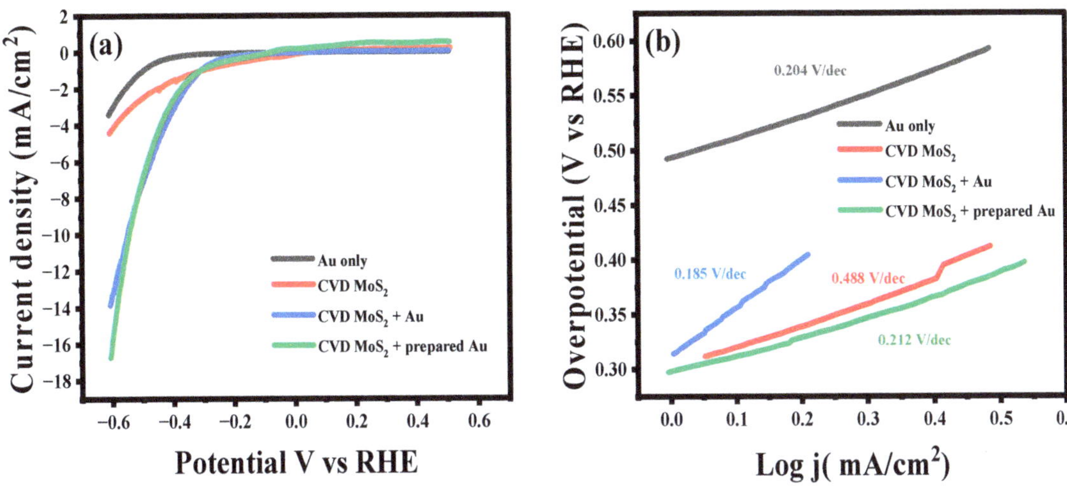

Figure 5. (**a**) The polarization curves of the Au, CVD-MoS_2, CVD-MoS_2 + Au, and CVD-MoS_2 + prepared Au. (**b**) The Tafel plot of the Au, CVD-MoS_2, CVD-MoS_2+ Au and CVD-MoS_2 + prepared Au composites in H_2SO_4.

The corresponding Tafel plots in Figure 5b and values in Table 3 support these observations. The Tafel slope is a key indicator of hydrogen evolution reaction (HER) kinetics, with lower values denoting more favorable catalytic performance. As shown, the pristine CVD-MoS_2 electrode exhibits a Tafel slope of 0.488 V/dec, while the hybrid catalysts show lower slopes of 0.212 V/dec for CVD-MoS_2 + prepared Au and 0.185 V/dec for CVD-MoS_2 + Au, indicating more favorable HER kinetics. Although the exchange current densities for the hybrids (0.199 and 0.0123 mA/cm^2, respectively) are lower than that of unmodified

MoS_2, the reduced Tafel slopes and higher current densities reflect an overall enhancement in HER activity. In contrast, the Au-only catalyst shows poor performance, with a high onset potential of 0.492 V, a cathodic current of only 1.15 mA/cm^2, a Tafel slope of 0.204 V/dec, and the lowest j_0 value (0.0027 mA/cm^2). These findings suggest that the controlled synthesis of hybrid materials combining Au and MoS_2 can significantly enhance HER efficiency by leveraging beneficial electronic interactions and robust performance across various environments.

Table 3. Electrochemical parameters were calculated to evaluate the HER performances of all the prepared MoS_2 catalysts in H_2SO_4.

Catalyst	Onset Potential (V)	Cathodic Current * (mA/cm^2)	Tafel (V/dec)	Exchange Current $mA\ cm^{-2}$ (j_0)
Au	0.492	1.15	0.204	0.0027
CVD-MoS_2	0.311	2.6	0.488	0.11
CVD-MoS_2 + prepared Au	0.304	7.48	0.212	0.199
CVD-MoS_2 + Au	0.304	7.27	0.185	0.0123

The enhanced HER activity observed in the Au-modified MoS_2 flakes can be directly correlated with their physicochemical characteristics. SEM imaging confirms the preservation of the flake-like morphology after Au nanoparticle deposition, while XRD patterns and Raman spectra indicate high crystallinity and a well-defined 2H-MoS_2 phase. This structural order is crucial for maintaining efficient charge carrier pathways. The introduction of Au nanoparticles provides additional active sites and improves electrical conductivity, which facilitates faster electron transfer, particularly at the MoS_2–Au interface. This is evidenced by the decreased onset potential and increased exchange current density. Furthermore, the Tafel slope of ~126 mV/dec suggests that the HER follows a Volmer–Heyrovsky mechanism, with the Heyrovsky step being the rate-limiting process. The presence of Au likely accelerates this step by enhancing surface electron availability, leading to improved overall kinetics. Thus, the synergistic effect between MoS_2 flakes and Au nanoparticles underpins the superior catalytic performance.

2.7. Chronoamperometry Studies of the MoS_2+ Au Sample

The stability levels of all prepared samples were evaluated using chronoamperometric measurements. To assess the stability of Au, CVD-MoS_2, CVD-MoS_2 + Au, and CVD-MoS_2 + prepared Au composites, chronoamperometry was conducted at a fixed potential of 0.5 V vs. RHE for up to 1000 min in neutral medium, as shown in Figure 6. The results demonstrate that the electrodes modified with Au, CVD-MoS_2, CVD-MoS_2 + Au, and CVD-MoS_2 + prepared Au composites exhibited stable and continuous current responses with minimal decline over time. The gradual decrease in current density observed in Figure 6 can be primarily attributed to the rapid diffusion of electroactive species at the electrode interface. After an initial current drop during the first 5 min, the current stabilized and remained nearly constant for the remainder of the test duration (up to 1000 minutes). This sustained current output highlights the electrochemical stability of the electrodes. Overall, the chronoamperometric data indicate that all tested composites effectively retained their initial current densities for up to 15 h, demonstrating consistent performance and high efficiency in promoting hydrogen evolution reaction (HER) activity following the initial decay phase. The constant and consistent current densities over a long period of time strongly indicate that any potential gold leaching was very small and negligible and did not affect the catalytic performance.

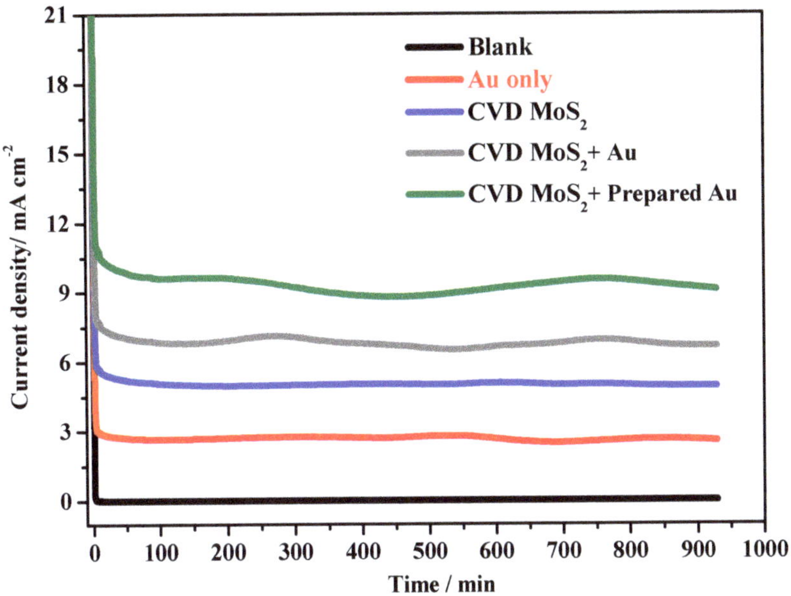

Figure 6. The chronoamperometry of the Au, CVD-MoS_2, CVD-MoS_2 + Au and CVD-MoS_2 + prepared Au composites in neutral medium at the applied potential of 0.5 V vs. RHE.

2.8. Comparative Study of Recent Literature on the Use of MoS_2/Au as a Catalyst for the HER

As presented in Table 4, the electrochemical performance of the synthesized composite was benchmarked against data from prior studies. While the observed Tafel slope exceeded the typical range reported for MoS_2-based catalysts, this discrepancy may have arisen from inherent material properties or synthesis-induced variations that impeded charge transfer kinetics. Despite this, the composite exhibited a markedly enhanced onset potential of 154 mV, which is substantially lower than previously reported values. This improvement suggests more favorable initial proton adsorption and earlier activation of the hydrogen evolution reaction (HER). Collectively, our results highlight the enhanced HER performance of the MoS_2/Au hybrid system, particularly in neutral media, and underscore the critical role of gold nanoparticle incorporation in improving both catalytic efficiency and operational durability. While gold is indeed a precious metal, its overall contribution to the cost of the MoS_2/Au catalyst is modest due to the low loading and high catalytic efficiency achieved. Notably, our catalyst exhibits superior electrocatalytic performance in a neutral medium, significantly reducing the actual cost per application. Furthermore, with continued developments in scalable synthesis technologies and gold nanoparticle extraction methods, the economic barriers associated with gold are expected to decrease, enhancing the feasibility of the MoS_2/Au system for broader practical applications.

Table 4. A comparison of recent literature on the use of MoS_2/Au as a catalyst for the HER.

Catalyst	Supporting Electrode	Electrolyte	HER onset (V)	Tafel (mV/dec)	Ref
MoS_2-Au	GC	H_2SO_4	0.2	50	[11]
MoS_2-Au	GC	KCl	0.27	71	[37]
MoS_2-Au	Si substrate	H_2SO_4	0.340	99.8	[12]
W- MoS_2-Au	Si substrate	H_2SO_4	0.190	86.9	[12]
CVD-MoS_2	GC	H_2SO_4	0.31	488	This work
CVD-MoS_2/Au	GC	H_2SO_4	0.304	185	This work
CVD-MoS_2/Prepared Au	GC	H_2SO_4	0.306	212	This work
CVD-MoS_2	GC	KCl	0.215	308	This work
CVD-MoS_2/Au	GC	KCl	0.152	255	This work
CVD-MoS_2/Prepared Au	GC	KCl	0.289	126	This work

3. Experimental Procedures

3.1. Materials and Reagent

The chemicals were of analytical grade and utilized as received without further purification. Hydrogen tetrachloroaurate (III) trihydrate ($HAuCl_4 \cdot 3H_2O$- Sigma Aldrich Ltd., St. Louis, MI, USA), Trisodium citrate ($C_6H_5Na_3O_7$- Sigma-Aldrich, St. Louis, MI, USA), MoO_3 (99.5% AlfaAesar, Heysham, UK), S powders (99.9%, VWR Chemicals, San Diego, CA, USA), and NaCl were employed. All solutions were prepared using double-distilled water at room temperature (25 °C).

3.2. Synthesis of Au

A total of 2.0 mL of trisodium citrate solution (0.01M) was added to a 250 mL round-bottom flask containing 50 mL of water and stirred with a magnetic stirrer. Upon reaching boiling point, 1 mL of the 0.01 M prepared $HAuCl_4$ solution was swiftly added to the boiling mixture. Over 5 min, the solution transitioned in color from light purple, indicative of larger particle sizes, to ruby red, signaling the formation of nano-sized particles. Following 10 min of boiling, the heating source was turned off, allowing the colloid to cool naturally [38–40].

3.3. Synthesis of CVD-MoS_2

A water solution containing sodium chloride (NaCl) and molybdenum trioxide (MoO_3) in a 40% ratio was prepared and subjected to sonication for 45 min. A small amount of this solution was dropped onto a quartz boat and heated to 323 K to remove the water content, leaving homogenous coverage. It was placed in a 1.5-inch quartz tube furnace positioned 18 cm from a 200 mg sulfur powder. Initially, the tube was cleansed by purging with a high flow rate of nitrogen gas (N_2) before reducing it to 5 SCCM to maintain the sublimated vapor until the growth temperature was reached. The temperature of the furnace was subsequently elevated to 780 °C. Meanwhile, sulfur was heated externally to 250 °C and introduced into the system at a flow rate of 20 standard cubic centimeters per minute (SCCM) for a 20-minute growth period. Then, the furnace cooled down naturally to reach room temperature. The resulting flakes material was collected from the boat in powdered form, which was then dispersed in distilled water and sonicated for 3 hours to yield a solution of 2D MoS_2 flakes [41].

CVD-MoS_2/Au: A total of 1 mL of suspension CVD-MoS_2 was added to a 10 mL glass bottle, then 1 mL of $HAuCl_4$ solution was gradually added during magnetic stirring. The mixture was ultrasonicated for 5 min to ensure uniformity. Subsequently, a volume of 3 mL of sodium citrate solution was introduced into the mixture, and the glass container was transferred to a water bath set at a temperature of 96 °C while being subjected to magnetic stirring. This process continued until a homogeneously distributed black–purple mixture was achieved [9,42].

CVD-MoS_2/prepared Au: A total of 1 mL of suspension CVD-MoS_2 was added into a 10 mL glass bottle, then 4 mL of prepared Au solution (mentioned previously) was gradually added during magnetic stirring. The mixture was ultrasonicated for 5 min to ensure uniformity. Then, it was transferred to a water bath set at a temperature of 96 °C while being subjected to magnetic stirring. This process continued until a homogeneously distributed black–purple mixture was achieved.

3.4. Fabrication of MoS_2 Modified GCE

The electrode was prepared using the following steps: First, 5 mL of CVD-MoS_2 solution and 30 µL of Nafion solution were mixed and dispersed at room temperature, followed by a 1-hour ultrasonic treatment. Then, 5 µL of the resulting dispersion was drop-cast onto the surface of a 3 mm glassy carbon electrode (GCE). The modified GCE was left to dry at room temperature for 24 h [43].

3.5. Samples Characterization

The samples were analyzed using FTIR spectroscopy. X-ray diffraction (XRD) analysis (Malvern, GH Eindhove, Netherlands), with Cu Kα radiation (λ = 0.154 nm), was conducted to assess crystallinity. Raman spectroscopy was performed with a Raman microscope operating at an excitation wavelength of 445 nm. The sample morphologies were characterized using scanning electron microscopy (SEM) (Tescan Vega 3 SBU, Czech Republic), and the energy-dispersive X-ray (EDX) spectra were obtained with the same SEM instrument. Dynamic light scattering (DLS) analysis was also performed to determine the size of the Au particles.

3.6. Electrochemical Measurements

All electrochemical measurements were conducted at room temperature in either 0.5 M H_2SO_4 or 1.0 M KCl solutions using a PGSTAT204 electrochemical potentiostat (Metrohm Autolab, Netherlands). A three-electrode system was used, where the fabricated electrode served as the working electrode, Ag/AgCl/KCl(sat.) served as the reference electrode, and a Pt wire served as the counter electrode. Linear sweep voltammetry (LSV) was performed in the potential range of −1 to 0.3 V at a scan rate of 0.1 V/s. Chronoamperometry measurements were conducted by applying a constant potential of +0.5 V versus a saturated Ag/AgCl reference electrode (Ag/AgCl/KCl(sat.)) using a standard three-electrode electrochemical cell configuration. The working electrode was held at this fixed potential throughout the experiment to monitor the resulting current response over time, allowing for the analysis of electrochemical kinetics and diffusion-controlled processes. All potential values were recorded for the Ag/AgCl reference electrode and then calibrated to the reversible hydrogen electrode (RHE).

4. Conclusions

In this study, we synthesize CVD-MoS_2 flakes, modify the sample with Au nanoparticles, and evaluate their performance for the hydrogen evolution reaction (HER). All the analysis tools, such as SEM, XRD, and Raman spectroscopy, confirm the 2D MoS_2 flake structure with high crystalline. The HER performance of the CVD-MoS_2 and Au-modified MoS_2 flakes is assessed through polarization techniques in both acidic and neutral electrolytes. The Au-MoS_2 flakes exhibit a significant reduction in onset potential (0.152 V at 1.0 mA/cm^2), a higher exchange current density (j_0 = 0.22 mA/cm^2), and a maximum current density of 8.4 mA/cm^2 at an overpotential of 0.372 V, demonstrating superior catalytic activity. These findings highlight the fact that integrating Au nanoparticles with MoS_2 can effectively enhance catalytic efficiency, offering a promising approach for efficient hydrogen production. The observed Tafel slopes (~126 mV/dec) suggest that the HER proceeds via a Volmer–Heyrovsky mechanism, wherein the initial proton adsorption (the Volmer step) is relatively fast and the electrochemical desorption of hydrogen (the Heyrovsky step) governs the overall kinetics. The enhanced performance upon Au integration indicates improved charge transfer at the MoS_2–Au interface, likely accelerating the Heyrovsky step. Chronoamperometry at 0.5 V vs. RHE for 1000 min shows that all electrodes maintain stable current with minimal decay, confirming their electrochemical stability and efficiency for sustained HER activity.

Author Contributions: Conceptualization, S.H.A., S.S.L., K.A.-A. and H.M.A.E.-L.; methodology, S.H.A. and K.A.-A.; formal analysis, S.H.A. and S.S.L.; investigation, K.A.-A., S.S.L., S.H.A. and H.M.A.E.-L.; writing—original draft preparation, S.H.A. and S.S.L.; review and editing, S.H.A. and H.M.A.E.-L.; supervision, K.A.-A. and S.H.A. All authors have read and agreed to the published version of the manuscript.

Funding: This work is funded by King Faisal University, Saudi Arabia [Project No. KFU251515].

Data Availability Statement: Data is contained within this article.

Acknowledgments: This work was supported by the Deanship of Scientific Research, Vice Presidency for Graduate Studies and Scientific Research, King Faisal University, Saudi Arabia [Grant No. KFU251515].

Conflicts of Interest: The authors declare no conflicts of interest.

References

1. Cho, S.C.; Seok, J.H.; Manh, H.N.; Seol, J.H.; Lee, C.H.; Lee, S.U. Expanding the frontiers of electrocatalysis: Advanced theoretical methods for water splitting. *Nano Convergence* **2025**, *12*, 4. [CrossRef] [PubMed]
2. Manyepedza, T.; Courtney, J.M.; Snowden, A.; Jones, C.R.; Rees, N.V. Impact Electrochemistry of MoS_2: Electrocatalysis and Hydrogen Generation at Low Overpotentials. *J. Phys. Chem. C.* **2022**, *126*, 17942–17951. [CrossRef]
3. Al-Amer, K.; Laradhi, S.S.; Aleithan, S.H.; Abd El-Lateef, H.M. Synthesis, Characterization, and Hydrogen Evolution Reaction Activity of MoS_2 Nanostructures Prepared Using Nonionic Surfactant. *Catalysts* **2024**, *14*, 804. [CrossRef]
4. Muralikrishna, S.; Manjunath, K.; Samrat, D.; Reddy, V.; Ramakrishnappa, T.; Nagaraju, D.H. Hydrothermal synthesis of 2D MoS_2 nanosheets for electrocatalytic hydrogen evolution reaction. *RSC Adv.* **2015**, *5*, 89389–89396. [CrossRef]
5. Zhao, X.; He, D.-W.; Wang, Y.-S.; Fu, C. In situ growth of different numbers of gold nanoparticles on MoS_2 with enhanced electrocatalytic activity for hydrogen evolution reaction. *Chinese Phys. B* **2018**, *27*, 068103. [CrossRef]
6. Chu, S.; Majumdar, A. Opportunities and challenges for a sustainable energy future. *Nature* **2012**, *488*, 294–303. [CrossRef]
7. Kalamaras, C.M.; Efstathiou, A.M. Hydrogen Production Technologies: Current State and Future Developments. *Conf. Papers Energy* **2013**, *3*, 1–9. [CrossRef]
8. Megía, P.J.; Vizcaíno, A.J.; Calles, J.A.; Carrero, A. Hydrogen Production Technologies: From Fossil Fuels toward Renewable Sources. A Mini Review. *Energy Fuels* **2021**, *35*, 16403–16415. [CrossRef]
9. Lu, K.; Liu, J.; Dai, X.; Zhao, L.; Yang, Y.; Li, H.; Jiang, Y. Construction of a Au@MoS_2 composite nanosheet biosensor for the ultrasensitive detection of a neurotransmitter and understanding of its mechanism based on DFT calculations. *RSC Adv.* **2022**, *12*, 798–809. [CrossRef]
10. Chen, B.; Hu, P.; Yang, F.; Hua, X.; Yang, F.F.; Zhu, F.; Sun, R.; Hao, K.; Wang, K.; Yin, Z. In Situ Porousized MoS_2 Nano Islands Enhance HER/OER Bifunctional Electrocatalysis. *Small* **2023**, *19*, 2207177. [CrossRef]

11. Bar-Ziv, R.; Ranjan, P.; Lavie, A.; Jain, A.; Garai, S.; Bar Hen, A.; Popovitz-Biro, R.; Tenne, R.; Arenal, R.; Ramasubramaniam, A.; et al. Au-MoS_2 Hybrids as Hydrogen Evolution Electrocatalysts ACS Appl. *Energy Mater.* **2019**, *2*, 6043–6050.
12. Li, Y.; Majewski, M.B.; Islam, S.M.; Hao, S.; Murthy, A.A.; DiStefano, J.G.; Hanson, E.D.; Xu, Y.; Wolverton, C.; Kanatzidis, M.G.; et al. Morphological Engineering of Winged Au@MoS_2 Heterostructures for Electrocatalytic Hydrogen Evolution. *Nano Lett.* **2018**, *18*, 7104–7110.
13. Zhou, Q.; Hu, H.; Chen, Z.; Ren, X.; Ma, D. Enhancing electrocatalytic hydrogen evolution via engineering unsaturated electronic structures in MoS_2. *Chem. Sci.* **2025**, *16*, 1597–1616. [CrossRef] [PubMed]
14. Obiang Nsang, G.E.; Ullah, B.; Hua, S.; Shah, S.A.; Ullah, N.; Ullah, N.; Dike, F.N.; Yaseen, W.; Yuan, A.; Khan, N.; et al. NiS nanoparticle-MoS_2 nanosheet core-shell spheres: PVP-assisted synthesis and efficient electrocatalyst for hydrogen evolution reaction. *Energy Mater.* **2025**, *5*, 500047. [CrossRef]
15. Jing, Y.; Wang, R.; Wang, Q.; Wang, X. Gold Nanoclusters Grown on MoS_2 Nanosheets by Pulsed Laser Deposition: An Enhanced Hydrogen Evolution Reaction. *Molecules* **2021**, *26*, 7503. [CrossRef]
16. Kim, J.; Byun, S.; Smith, A.J.; Yu, J.; Huang, J. Enhanced Electrocatalytic Properties of Transition-Metal Dichalcogenides Sheets by Spontaneous Gold Nanoparticle Decoration. *J. Phys. Chem. Lett.* **2013**, *4*, 1227–1232. [CrossRef]
17. Zhao, S.; Jin, R.; Song, Y.; Zhang, H.; House, S.D.; Yang, J.C.; Jin, R. Atomically Precise Gold Nanoclusters Accelerate Hydrogen Evolution over MoS_2 Nanosheets: The Dual Interfacial Effect. *Small* **2017**, *13*, 1701519. [CrossRef]
18. Shakya, J.; Patel, A.S.; Singh, F.; Mohanty, T. Composition dependent Fermi level shifting of Au decorated MoS_2 nanosheets. *Applied Physics Letters.* **2016**, *108*, 013103. [CrossRef]
19. Zhang, J.; Wang, T.; Liu, L.; Du, K.; Liu, W.; Zhu, Z.; Li, M. Molybdenum disulfide and Au ultrasmall nanohybrids as highly active electrocatalysts for hydrogen evolution reaction. *J. Mater. Chem. A.* **2017**, *5*, 4122–4128. [CrossRef]
20. Shi, Y.; Huang, J.-K.; Jin, L.; Hsu, Y.-T.; Yu, S.F.; Li, L.-J.; Yang, H.Y. Selective Decoration of Au Nanoparticles on Monolayer MoS_2 Single Crystals. *Sci. Rep.* **2013**, *3*, 1839. [CrossRef]
21. Zhao, J.; Zhang, Z.; Yang, S.; Zheng, H.; Li, Y. Facile synthesis of MoS_2 nanosheet-silver nanoparticles composite for surface enhanced Raman scattering and electrochemical activity. *J. Alloys Compd.* **2013**, *559*, 87–91. [CrossRef]
22. Shomalian, K.; Bagheri-Mohagheghim, M.-M.; Ardyanian, M. Characterization and study of reduction and sulfurization processing in phase transition from molybdenum oxide (MoO2) to molybdenum disulfide (MoS_2) chalcogenide semiconductor nanoparticles prepared by one-stage chemical reduction method. *Appl. Phys. A.* **2017**, *123*, 93. [CrossRef]
23. Ali, G.A.M.; Thalji, M.R.; Soh, W.C.; Algarni, H.; Chong, K.F. One-step electrochemical synthesis of MoS_2/graphene composite for supercapacitor application. *J. Solid State Electrochem.* **2020**, *24*, 25–34. [CrossRef]
24. Venkateshwaran, S.; Senthil Kumar, S.M. Provoking Metallic 1T Phase Conversion of 2H-MoS_2 via an Effectual Solvothermal Route for Electrocatalytic Water Reduction in Acid. *ACS Sustainable Chem. Eng.* **2022**, *10*, 5258–5267. [CrossRef]
25. Saadati, M.; Akhavan, O.; Fazli, H. Single-Layer MoS_2-MoO3-x Heterojunction Nanosheets with Simultaneous Photoluminescence and Co-Photocatalytic Features. *Catalysts* **2021**, *11*, 1445. [CrossRef]
26. Shan, X.; Zhang, S.; Zhang, N.; Chen, Y.; Gao, H.; Zhang, X. Synthesis and characterization of three-dimensional MoS_2@carbon fibers hierarchical architecture with high capacity and high mass loading for Li-ion batteries. *J. Colloid Interface Sci.* **2018**, *510*, 327–333. [CrossRef]
27. Nagarajan, T.; Khalid, M.; Sridewi, N.; Jagadish, P.; Shahabuddin, S.; Muthoosamy, K.; Walvekar, R. Tribological, oxidation and thermal conductivity studies of microwave synthesised molybdenum disulfide (MoS_2) nanoparticles as nano-additives in diesel based engine oil. *Sci. Rep.* **2022**, *12*, 14108. [CrossRef]
28. Hasmin, H.F.; Imawan, C.; Fauzia, V. The Role of Temperature in the Hydrothermal Synthesis on the Structural and Morphological Properties of MoS_2. *J. Phys. Conf. Ser.* **2021**, *1951*, 012014.
29. Zhang, J.; Han, D.; Wang, Y.; Wang, L.; Chen, X.; Qiao, X.; Yu, X. Synergy between nanozymes and natural enzymes on the hybrid MoS_2 nanosheets/graphite microfiber for enhanced voltammetric determination of hydrogen peroxide. *Microchim Acta* **2020**, *187*, 321. [CrossRef]
30. Chen, Q.; Wang, K.; Qin, J.; Wang, S.; Wei, W.; Wang, J.; Shen, Q.; Qu, P.; Liu, D. Rux Se@MoS_2 hybrid as a highly efficient electrocatalyst toward hydrogen evolution reaction. *RSC Adv.* **2019**, *9*, 13486–13493. [CrossRef]
31. Biswas, S.; Ranwa, S. CVD-Grown MoS_2 Nanosheets-Based Gas Sensor for Low-Limit Detection of NO_2 Gas. In *Proceedings of the 4th International Conference on Communication, Devices and Computing*; Lecture Notes in Electrical Engineering; Sarkar, D.K., Sadhu, P.K., Bhunia, S., Samanta, J., Paul, S., Eds.; Springer: Singapore, 2023; Volume 1046, pp. 439–447.
32. Thangappan, R.; Dhinesh Kumar, R.; Jayavel, R. Facile Synthesis and Characterization of Molybdenum Oxide (MoO_3) Nanofibers and Submicron Rods by Electrospinning Technique for Potential Application in Photocatalytic Activity. *J. Clust Sci.* **2022**, *33*, 2209–2214. [CrossRef]
33. Solanki, K.N.; Cadambi, R.M.; Srikari, S.; Shankapal, S.R. Synthesis of Nanostructured Molybdenum Disulfide by Chemical Vapour Deposition. *Mater. Today Proc.* **2017**, *4*, 11134–11140. [CrossRef]

34. Ghahramanifard, F.; Rouhollahi, A.; Fazlolahzadeh, O. Synthesis of n-type Cu-doped ZnO Nanorods onto FTO by Electrodeposition Method and Study its Electrocatalytic Properties toward CO_2 Reduction. *Anal. Bioanal. Electrochem.* **2018**, *10*, 362–371.
35. Singh, S.; Deb, J.; Sarkar, U.; Sharma, S. MoS_2/MoO_3 Nanocomposite for Selective NH_3 Detection in a Humid Environment. *ACS Sustain. Chem. Eng.* **2021**, *9*, 7328–7340. [CrossRef]
36. Aleithan, S.H.; Al-Amer, K.; Alabbad, Z.H.; Khalaf, M.M.; Alam, K.; Alhashem, Z.; Abd El-Lateef, H.M. Highly scalable synthesis of MoS_2 thin films for carbon steel coatings: Influence of synthetic route on the nanostructure and corrosion performance. *J. Mater. Res. Technol.* **2023**, *23*, 1239–1251. [CrossRef]
37. Shi, Y.; Wang, J.; Wang, C.; Zhai, T.-T.; Bao, W.-J.; Xu, J.-J.; Xia, X.-H.; Chen, H.-Y. Hot Electron of Au Nanorods Activates the Electrocatalysis of Hydrogen Evolution on MoS_2 Nanosheets. *J. Am. Chem. Soc.* **2015**, *137*, 7365–7370. [CrossRef]
38. Dong, J.; Carpinone, P.L.; Pyrgiotakis, G.; Demokritou, P.; Moudgil, B.M. Synthesis of Precision Gold Nanoparticles Using Turkevich Method. *KONA* **2020**, *37*, 224–232. [CrossRef] [PubMed]
39. Hussain, M.H.; Abu Bakar, N.F.; Mustapa, A.N.; Low, K.-F.; Othman, N.H.; Adam, F. Synthesis of Various Size Gold Nanoparticles by Chemical Reduction Method with Different Solvent Polarity. *Nanoscale Res. Lett.* **2020**, *15*, 140. [CrossRef]
40. Herizchi, R.; Abbasi, E.; Milani, M.; Akbarzadeh, A. Current methods for synthesis of gold nanoparticles. *Artif. Cells Nanomed. Biotechnol.* **2016**, *44*, 596–602. [CrossRef]
41. Aleithan, S.H.; Al-Amer, K.; Alhashem, Z.; Alati, N.A.; Alabbad, Z.H.; Alam, K. Growth of MoS_2 films: High-quality monolayered and multilayered material. *AIP Adv.* **2022**, *12*, 075220. [CrossRef]
42. Wang, J.; Zhang, W.; Chen, L.; Huang, J.; Shi, X.; Han, D.; Wen, J.; Yan, H. Preparation of MoS_2 @AuNP nanocomposite by a self-reduction method and its application for electrochemical glucose sensing. *Mater. Adv.* **2022**, *3*, 8677–8683. [CrossRef]
43. Zeng, X.; Niu, L.; Song, L.; Wang, X.; Shi, X.; Yan, J. Effect of Polymer Addition on the Structure and Hydrogen Evolution Reaction Property of Nanoflower-Like Molybdenum Disulfide. *Metals* **2015**, *5*, 1829–1844. [CrossRef]

catalysts

MDPI

Article

Post-Synthesis Ion Beam Sputtering of $Pt/CeO_2–ZrO_2$ Catalysts: Correlating Surface Modifications with Light-Off Performance

Ruairi O'Donnell [1], Marina Maddaloni [2,3], Salvatore Scaglione [4] and Nancy Artioli [2,3,*]

1 School of Chemistry and Chemical Engineering, Queen's University Belfast, David-Keir Building, Stranmillis Road, Belfast BT9 5AG, UK; rodonnell17@qub.ac.uk
2 CEEP Laboratory, Department of Civil Engineering, Architecture, Territory, Environment and Mathematics, University of Brescia, Via Branze 38, 25123 Brescia, Italy; marina.maddaloni@unibs.it
3 Consorzio Interuniversitario Nazionale per la Scienza e Tecnologia dei Materiali (INSTM), University of Brescia, Via Branze 38, 25123 Brescia, Italy
4 ENEA-CRE Casaccia, Via Anguillaiese 301, 00060 Roma, Italy; salvatore.scaglione@enea.it
* Correspondence: nancy.artioli@unibs.it

Abstract: High-efficiency diesel and lean-burn engines produce lower exhaust temperatures, which can delay the activation of after-treatment catalysts such as Diesel Oxidation Catalysts (DOCs). This study explores ion beam sputtering as a post-synthesis strategy to enhance the low-temperature activity of commercial $Pt/CeO_2–ZrO_2$ catalysts. Low-energy ions (0.5–1.5 keV) were applied with controlled variations in treatment number, beam current, and exposure time to selectively generate oxygen vacancies and improve Pt dispersion. Structural and chemical effects were characterized using X-ray diffraction (XRD), BET surface area measurements, X-ray photoelectron spectroscopy (XPS) and extended X-ray absorption fine structure (EXAFS). Catalytic performance was evaluated through CO and C_3H_6 oxidation under conditions mimicking lean-burn engine exhaust. Increasing the number of ion treatments progressively lowered light-off temperatures, correlating with enhanced Pt–Ce^{3+} interactions and improved surface reducibility. Variations in beam current and exposure time further modulated these surface effects, confirming the tunable nature of the approach. The results demonstrate that ion beam sputtering selectively modifies the catalyst surface without altering the bulk structure, directly linking atomic-scale modifications to improved low-temperature activity. This strategy offers a promising route to overcome delayed light-off issues in modern high-efficiency engines, providing a precise, controllable method to optimize emission control catalysts.

Keywords: ion beam sputtering; $Pt/CeO_2–ZrO_2$ catalysts; automotive emission control; light-off temperature; CO and hydrocarbon oxidation; post-synthesis surface modification

1. Introduction

Since the early 1990s, the European Commission's Thematic Strategy on Air Pollution has progressively introduced more stringent standards to reduce transport-related emissions (CO, hydrocarbons, NOx, PM, etc.), as part of a broader effort to improve air quality and, more broadly, to support the EU's long-term objective of achieving carbon neutrality by 2050, in line with ongoing research and strategies for sustainable energy and low-carbon technologies [1–4]. The transition from Euro 5 in 2009 to Euro 6/VI in 2014 marked a significant tightening of NOx emission limits for diesel engines: from 0.18 g/km (Euro 5) to 0.08 g/km (Euro 6) for light-duty vehicles and from 2.0 g/kWh (Euro V) to 0.4 g/kWh (Euro VI) for heavy-duty vehicles. This regulatory change accelerated the adoption of advanced after-treatment technologies, including Selective Catalytic Reduction

(SCR), Lean NOx Traps (LNT), and Diesel Oxidation Catalysts (DOC). Building on this trajectory, the forthcoming Euro 7 regulation—scheduled to apply from 2027—further strengthens emission limits across all vehicle categories. Unlike previous standards, Euro 7 introduces a unified framework for light- and heavy-duty vehicles, expands testing conditions (e.g., real-world driving emissions over a wider temperature and lifetime range), and places stricter durability requirements on emission-control systems. These measures are expected to drive the next generation of catalyst technologies and integrated powertrain solutions [1,5,6].

Many of these catalytic converters, particularly DOCs, are based on Pt–ceria/zirconia materials, where the complementary functionalities of ceria and Pt play a pivotal role. In particular, the incorporation of ZrO_2 into CeO_2 has been widely recognized as a strategy to enhance both the structural and functional properties of ceria-based catalysts. Zirconium addition improves the thermal stability of the mixed oxide, reducing the propensity for sintering at high operating temperatures, which is critical for maintaining the dispersion of noble metal nanoparticles under harsh exhaust conditions. Moreover, the formation of CeO_2–ZrO_2 solid solutions enhances the oxygen storage capacity (OSC) of the support, a key property for dynamic oxygen buffering during rich–lean cycling, which in turn facilitates the reversible Ce^{4+}/Ce^{3+} redox transition. This increase in OSC not only improves the redox flexibility of the support but also provides a greater density of surface defects and oxygen vacancies, which act as preferential anchoring sites for Pt atoms. By stabilizing Pt species and preventing their migration or sintering, the addition of ZrO_2 effectively supports higher metal dispersion and stronger metal–support interactions, thus contributing to the overall catalytic efficiency and durability of the system [7,8]. These advantages explain why Pt–ceria/zirconia catalysts remain the benchmark for modern automotive DOCs and provide the rationale for their widespread use in emission-control applications.

The remarkable performance of ceria arises from its redox flexibility, enabled by the facile and reversible transition between Ce^{4+} and Ce^{3+}, which confers a substantial oxygen storage capacity (OSC). This dynamic oxygen buffering, facilitated by the formation of oxygen vacancies, allows ceria to repeatedly adsorb and release O^{2-}, thereby promoting the oxidation of exhaust gases under transient rich–lean conditions [9]. At the same time, Pt provides exceptionally high oxidation activity, particularly toward NOx. When combined, the oxygen-buffering capacity of ceria-based supports and the intrinsic oxidation activity of Pt create a synergistic system that underpins the effectiveness of modern automotive emission-control catalysts [10].

A critical performance parameter of Diesel Oxidation Catalysts (DOCs) is the so-called light-off temperature, defined as the temperature at which catalytic reactions begin and, in practice, as the catalyst inlet temperature at which 50% of a given pollutant species is converted. Below this threshold, catalytic activity is minimal, while above it, the reaction rate increases exponentially with temperature. Achieving rapid light-off is therefore essential to ensure efficient pollutant conversion, particularly during the cold-start phase, when emissions are most severe [11,12].

The challenge, however, lies in the fact that modern diesel and lean-burn engines are designed for high thermal efficiency, which inherently results in lower exhaust gas temperatures. While this trend improves fuel economy and reduces CO_2 emissions, it simultaneously delays—or even prevents—after-treatment systems from reaching their activation threshold under certain operating conditions [13,14]. Technologies such as DOCs, Diesel Particulate Filters (DPFs), Lean NOx Traps (LNTs), and Selective Catalytic Reduction (SCR) all depend on surpassing their respective light-off temperatures to function effectively. During cold-start and low-load operation, however, the exhaust often remains below these critical temperatures, leaving the catalysts largely inactive and causing substantial emissions of NOx and particulate matter [15].

This limitation has become particularly critical under Euro 6/VI regulations, which enforce stringent emission limits not only under controlled laboratory tests but also in real driving conditions. As a result, significant research efforts are currently directed toward lowering the light-off temperature of catalytic materials or developing strategies to accelerate catalyst heating, thereby ensuring compliance with these demanding standards while preserving engine efficiency.

In this context, interface engineering using ion beam techniques has emerged as a powerful approach to tailor catalytic properties at the atomic scale, enabling precise structural modifications that can directly influence the catalyst's light-off temperature [16]. Among the different strategies, ion beam sputtering has attracted growing interest as a versatile tool to modify the nanoscale surface structure of advanced catalytic materials, particularly in the development of automotive after-treatment systems [17,18]. The use of low-energy accelerated ions (0.5–10 keV) enables controlled interactions with surface atoms of the catalyst without penetrating deeply into the bulk [19]. This process results in a "non-reactive removal" of atoms—commonly referred to as sputtering—where surface atoms are displaced without undergoing chemical reactions with the incident ions [20].

Beyond these general advantages, the selectivity of ion-beam sputtering can be particularly beneficial in mixed Pt–CeO_2 oxides. In this case, oxygen can be sputtered more readily than either Pt or ceria, thereby generating oxygen vacancies (V_O). These defects increase the concentration of Ce^{3+} in the structure, making the surface more reducible and receptive toward Pt, ultimately leading to stronger Pt–support interactions [21,22]. Several studies in the literature [22–24] have demonstrated that the creation of surface defects, including oxygen vacancies (V_O), can act as "energetic traps" for Pt atoms. Owing to the reducing nature of Ce^{3+}, the oxidation state of Pt can be modulated, which stabilizes Pt atoms on the surface, prevents their migration and sintering, and thus enhances their dispersion across the support [25].

Nevertheless, as with any emerging technique, some critical issues have also been raised. Wolf et al. [26] pointed out a potential limitation of ion sputtering and ion implantation for metal deposition: the reduction in the surface-active area caused by atoms being buried beneath the substrate. This observation underscores the need to carefully optimize ion beam parameters in order to exploit the advantages of the technique without compromising catalytic efficiency.

Yet, these challenges also highlight an underexplored opportunity. Only a limited number of studies have investigated the use of ion beams as a post-synthesis modification strategy, and here ion beam sputtering offers a distinctive advantage over other ion beam methods: it allows selective tuning of surface properties while leaving the bulk structure largely unaffected. This unique capability opens the way to targeted improvements in catalytic activity and stability. For example, Machida et al. [27] demonstrated the potential of post-synthesis ion implantation by employing O^+ and N^+ ions to modify noble metal catalysts supported on Al_2O_3 for methane oxidation, thereby providing experimental evidence of the effectiveness of ion-based surface processing.

In this study, we explore ion beam sputtering as a post-synthesis modification of commercial Pt/CeO_2–ZrO_2 catalysts for automotive emission control. Using low-energy ions, the surface is selectively modified to generate oxygen vacancies and enhance Pt dispersion, promoting a strong synergy between Pt and the redox-active ceria. Surface changes were characterized by XPS and EXAFS and complemented with SRIM simulations to elucidate the structural effects induced by sputtering.

Catalytic performance was evaluated via CO and hydrocarbon oxidation under conditions mimicking lean-burn engine exhaust. The enhanced interaction between Pt and Ce^{3+}–rich sites facilitates pollutant activation at lower temperatures, effectively addressing the delayed light-off issue associated with modern high-efficiency engines.

This work demonstrates that ion beam sputtering is an effective post-treatment strategy to tune catalyst surfaces, directly linking atomic-scale modifications to improved low-temperature activity in real-world emission control applications.

2. Results and Discussion

The structural stability of the $Pt/Ce_{0.5}Zr_{0.5}O_2$ catalyst after ion beam treatment was first assessed by X-ray diffraction (XRD) (Figure 1). All irradiated samples exhibited diffraction patterns consistent with the cubic fluorite-type $Ce_{0.5}Zr_{0.5}O_2$ solid solution, showing reflections at 2θ = 29.2°, 33.8°, 48.6°, 57.1°, 60.5°, 71.1°, and 78.7°, corresponding to the (111), (200), (220), (311), (222), (400), and (331) planes, respectively [28]. No appreciable changes in peak positions, intensities, or widths were observed, even in samples exposed to the most severe bombardment conditions (Pt1.5(16 × 4)), which accumulated 64 irradiation cycles. These results clearly indicate that ion beam sputtering does not compromise the long-range crystallinity of the $Ce_{0.5}Zr_{0.5}O_2$ lattice, supporting the hypothesis that structural modifications are confined to the near-surface region.

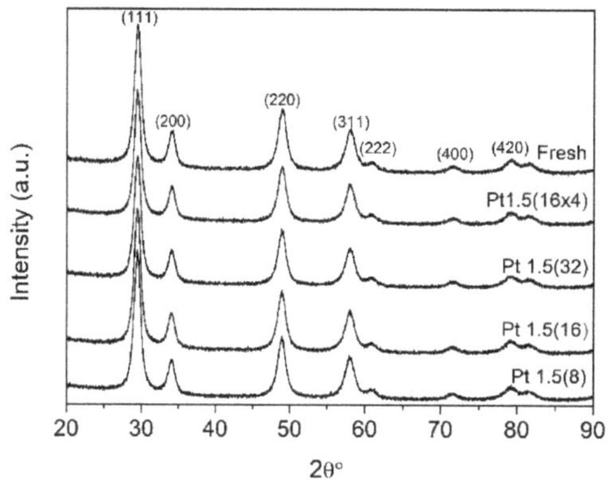

Figure 1. XRD patterns of the fresh 1 wt.% $Pt/Ce_{0.5}Zr_{0.5}O_2$ catalyst compared with those of the ion-bombarded samples. The number of irradiation treatments ranged from 8 cycles for Pt1.5(8) to 64 cumulative cycles for Pt1.5(16 × 4).

BET N_2 physisorption measurements further corroborated the bulk stability of the catalysts (Table 1). The fresh sample displayed a specific surface area of 88.2 $m^2\ g^{-1}$, and all treated materials retained values in the range of 86.7–89.9 $m^2\ g^{-1}$, well within the experimental error. This invariance confirms that ion bombardment does not affect textural properties such as porosity or accessible surface area. Together, the XRD and BET data establish that ion sputtering introduces subtle modifications localized at the surface without altering the overall framework of the mixed-oxide support.

Importantly, no Pt reflections were detected in any of the XRD patterns, as expected given the low noble metal loading (1 wt.%) and the high dispersion of Pt species, which lie below the detection limit of conventional diffractometry. Although XRD alone cannot provide direct evidence of enhanced dispersion upon ion treatment, complementary spectroscopic and structural data—including the downward shift of the PtO_x reduction peak in TPR, the decreased accessible Pt signal in XPS, and the reduction of surface Pt measured by XRF—indicate stronger metal–support interaction and redistribution of Pt species. This interpretation is further supported by our previous study on N^+-irradiated $Pt/Ce_{0.68}Zr_{0.32}O_2$ [16], where high-resolution TEM revealed a finer Pt particle distribution and partial embedding into a defect-rich oxide surface, providing direct microscopic confirmation that ion irradiation can stabilize highly dispersed Pt species. Together, these observations suggest that the current Ar^+ treatment induces similar defect-mediated redistribution and anchoring of Pt nanoparticles. This observation is also consistent with prior reports on Pt/ceria–zirconia materials [29], in which Pt is generally present as highly dispersed oxidic clusters or atomically anchored species rather than crystalline nanoparticles detectable by XRD. The combination of these findings strengthens the conclusion that the observed decrease in surface Pt measured by XRF does not imply detrimental loss of catalytic functionality, but rather a reorganization toward more active Pt sites.

Table 1. Ion beam treatment parameters applied to the post-synthesis modification of the fresh $Pt/Ce_{0.5}Zr_{0.5}O_2$ catalyst (1 wt.% Pt).

Sample	Energy keV	Current (mA)	No. of Treatments	Duration of Each Treatment (mins)	BET Surface Area (m^2 g^{-1})
Fresh	-	-	-	-	88.2
Pt1.5(16 × 4)	1.5	20	16 × 4 100 mg batches	15	87.3
Pt1.5(32)	1.5	20	32	15	86.8
Pt1.5(16)	1.5	20	16	15	86.7
Pt1.5(8)	1.5	20	8	15	89.7
Pt1.2(16)	1.2	20	16	15	88.2
Pt1.5(8)-10	1.5	10	8	15	87.2
Pt1.5(16)-2	1.5	20	16	2	89.9

While the bulk remained unchanged, temperature-programmed reduction (TPR) experiments revealed significant modifications in redox properties induced by ion sputtering (Figure 2). The fresh catalyst exhibited a well-defined H_2 consumption peak centered at ~415 K, attributable to the reduction of surface PtO_x species [30,31]. Given that the reduction of $Ce_{0.5}Zr_{0.5}O_2$ occurs above 773 K, no support-related features were detected in the investigated temperature range. Upon ion bombardment, this reduction peak progressively shifted toward lower temperatures, with onsets as low as 395 K. This indicates that Pt species became more easily reducible after irradiation, suggesting a modified electronic environment and stronger coupling with the support. Interestingly, the integrated H_2 uptake decreased systematically from 61.3 μmol g^{-1} in the fresh sample to 41–46 μmol g^{-1} in treated catalysts. This reduced hydrogen consumption is consistent with partial removal or redistribution of Pt during sputtering. Indeed, XRF analysis confirmed a measurable decrease in Pt loading at the surface after irradiation. Thus, although a fraction of Pt is lost, the remaining species become more reducible and more strongly interacting with the modified support. These TPR results are fully consistent with our recent findings on N^+-irradiated $Pt/Ce_{0.68}Zr_{0.32}O_2$ catalysts [16], where a comparable downward shift (~180 K) of the PtO_x reduction peak was observed. In that study, operando FTIR spectra in the νOH region provided direct spectroscopic evidence of oxygen vacancy formation, confirming that ion irradiation promotes selective oxygen removal and defect generation within the ceria–zirconia lattice. The parallel behavior observed here under Ar^+ sputtering further supports the conclusion that the enhanced reducibility originates from vacancy-mediated activation of the $Pt–CeO_2$ interfacial redox couple, rather than from a simple weakening of metal–support interactions.

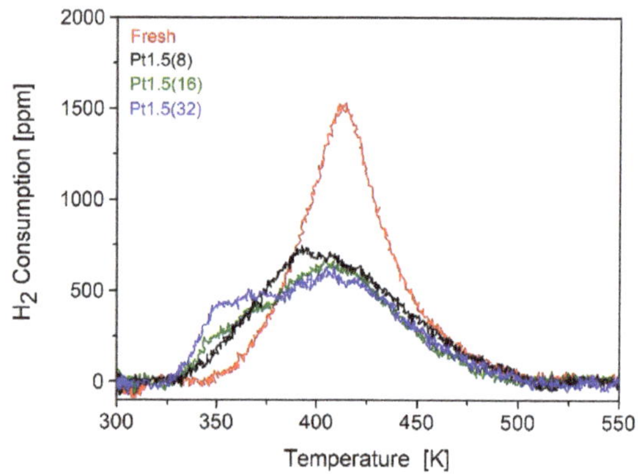

Figure 2. TPR profiles of the fresh 1 wt.% $Pt/Ce_{0.5}Zr_{0.5}O_2$ catalyst compared with those of the ion-bombarded samples, treated with increasing numbers of irradiation cycles ranging from 8 [Pt1.5(8)] to 32 [Pt1.5(32)].

X-ray photoelectron spectroscopy (XPS) provided further evidence of these changes (Figure 3). The Pt 4f spectra (Figure 3a) of all samples revealed predominantly oxidic Pt species, with Pt^{2+} as the dominant component. However, in the case of prolonged irradiation (Pt1.5(32)), a small fraction of metallic Pt^0 emerged, suggesting that extensive sputtering can locally alter the oxidation state of Pt. Moreover, overall Pt content (Figure 3b) decreased with increasing irradiation cycles, consistent with either partial Pt removal during sputtering or the formation of oxidic overlayers that attenuate the XPS signal [30]. More revealing were the modifications in the Ce 3d region (Figure 3c). Curve fitting of the multiplets, following the methodology of Romeo et al. [32], indicated a progressive increase in the fraction of Ce^{3+} with increasing ion exposure (Figure 3d). This trend aligns closely with the generation of oxygen vacancy (V_O), consistent with the expected preferential sputtering of oxygen atoms from the ceria lattice. A similar evolution of the Ce 3d spectra was also observed in our previously published work on N^+-irradiated $Pt/Ce_{0.68}Zr_{0.32}O_2$, where an increased Ce^{3+} contribution was attributed to partial lattice reduction and the formation of Ce^{3+}–V_O pairs at the surface. These defect sites were shown to act as preferential anchoring centers for Pt species, facilitating stronger metal–support coupling and enhancing interfacial redox flexibility. This interpretation is in excellent agreement with the mechanism proposed by Clark et al. [21], who reported that ion bombardment in Pt/CeO_2 systems generates surface oxygen vacancies that strengthen interfacial interactions. Therefore, the consistent evolution of both TPR and XPS features in our Ar^+- and N^+-irradiated catalysts strongly supports the conclusion that ion-beam-induced defect formation represents a general and reproducible mechanism to modulate the electronic and redox behavior of ceria–zirconia-supported systems. The formation of Ce^{3+}–V_O pairs is known to increase the reducibility of ceria and to provide anchoring sites for noble metals, thereby strengthening metal–support interactions. Such defect engineering has been widely reported as a key factor in enhancing the catalytic activity and stability of ceria-based materials [21,33,34].

The preferential generation of oxygen vacancies was further corroborated by SRIM-2013 simulations (Figure 4). At ion energies of 1.5 keV, the simulations predict that nearly 78% of sputtered atoms correspond to oxygen, a value significantly higher than the 66.7% expected from the stoichiometry of $Ce_{0.7}Zr_{0.3}O_2$. This selective oxygen removal, compared with only ~16–17% Ce and ~5% Zr, explains the enrichment in Ce^{3+} observed by XPS and the enhanced reducibility evidenced in TPR profiles. Importantly, the simulations confirm that oxygen depletion is not random but preferential, directly linking ion bombardment with the creation of catalytically relevant surface defects. This preferential vacancy formation provides the structural basis for the observed stabilization of Pt species and the enhanced interfacial redox activity, consistent with mechanisms involving hydrogen spillover and Ce^{3+}–V_O pairing described in Clark et al. [21] and supported by our previous N^+-irradiation study [16].

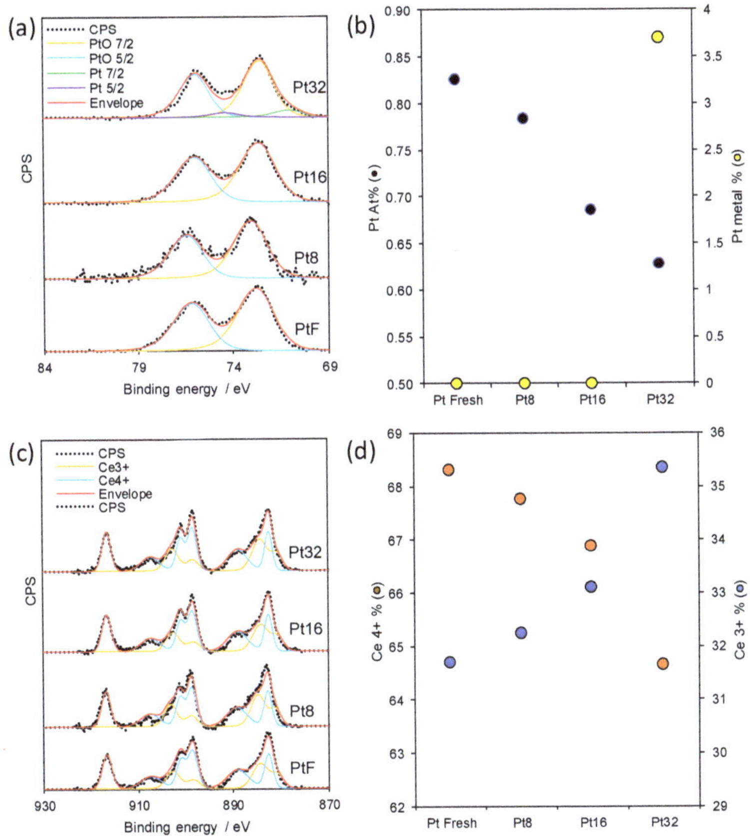

Figure 3. (**a**) Pt 4f XP spectra, (**b**) At% of Pt and deconvolution quantification of Pt metal content, (**c**) Ce 4f XP spectra, and (**d**) quantification of Ce^{3+} and Ce^{4+} in Pt Fresh, Pt8, Pt16, and Pt32 materials.

To confirm that these modifications were confined to ceria, the Zr 3d spectra were also examined (Figure 5).

Across all samples, zirconium remained fully oxidized as Zr^{4+}, unaffected by irradiation. This observation reinforces that ion sputtering selectively targets oxygen atoms in the ceria component while leaving the zirconia lattice chemically unaltered, in agreement with its lower sputtering yield.

The catalytic consequences of these surface modifications were evaluated under simulated lean-burn exhaust conditions. CO and C_3H_6 light-off curves are presented in Figure 6A and Figure 6B, respectively. The fresh catalyst exhibited T_{50} values of 513 K (CO) and 546 K (C_3H_6). Upon ion treatment, both light-off temperatures decreased progressively with the severity of irradiation. The most extensively bombarded catalyst, Pt1.5(16 × 4), achieved T_{50} values of 500 K for CO and 524 K for C_3H_6, representing improvements of 13 K and 22 K, respectively. Although modest in absolute numbers, these shifts are highly relevant in the context of automotive emission control, where lowering the light-off temperature by even a few tens of degrees can markedly reduce cold-start emissions. The correlation between increased Ce^{3+}–V_O populations and decreased light-off temperatures supports the mechanistic hypothesis that these vacancies enhance Pt–ceria interfacial redox synergy, promoting more efficient CO and hydrocarbon oxidation.

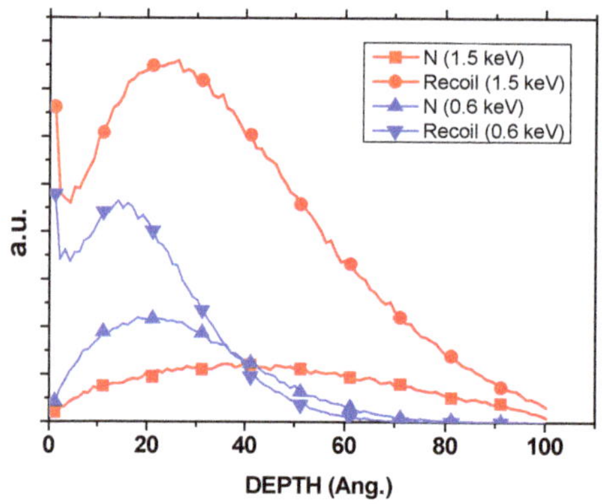

Figure 4. SRIM depth distribution profiles of incident N^+ ions and recoiled atoms within a 10 nm $Ce_{0.7}Zr_{0.3}O_2$ layer for two ion energies, 1.5 keV and 0.6 keV.

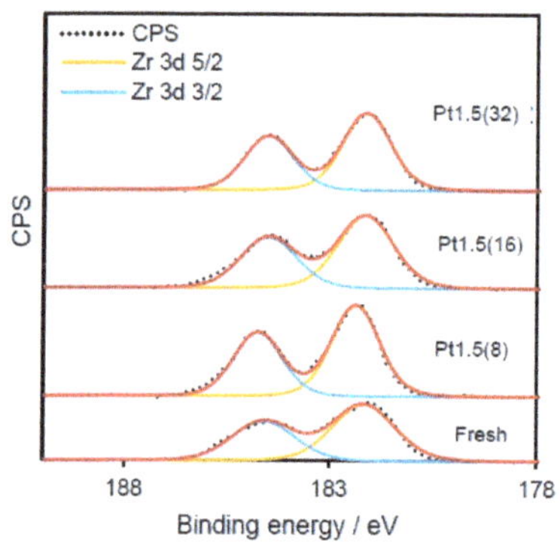

Figure 5. Zr 3d XPS spectra of Pt fresh, Pt8, Pt16, and Pt32 Pt/CeO_2–ZrO_2 catalysts after ion beam treatment. Red line is the fitted profile of Zr 3d XPS spectra.

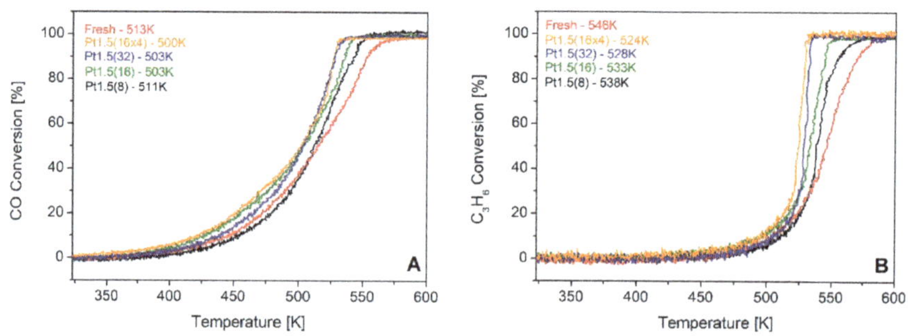

Figure 6. CO (**A**) and C_3H_6 (**B**) light-off curves obtained over a fresh Pt 1 wt.% $Ce_{0.5}Zr_{0.5}O_2$ catalyst compared to those of the samples that had been modified by ion bombardment. The number of treatments given to the samples ranges from 8 (Pt1.5(8)) to 64 (Pt1.5(16 × 4)).

The enhanced performance directly correlates with the increase in Ce^{3+} concentration and the improved reducibility observed by XPS and TPR, confirming that oxygen vacancies generated by sputtering enhance the activation of CO and hydrocarbons at lower temperatures. Interestingly, CH_4 oxidation remained largely unaffected by ion treatment, with T_{50} values unchanged relative to the fresh catalyst. This behavior is consistent with the inherently higher activation barrier for methane oxidation [27], which requires more drastic modifications (e.g., higher Pt loadings or different promoters) to be significantly improved.

To gain mechanistic insight, the influence of beam energy, current, and exposure time was systematically investigated. Variation of the accelerating voltage between 1.2 and 1.5 keV (Figure 7) produced essentially identical CO and C_3H_6 light-off temperatures, indicating that within this narrow energy window, the penetration depth of the ions does not play a dominant role. Similarly, reducing the beam current from 20 to 10 mA (Pt1.5(8) vs. Pt1.5(8)-10, Figure 8) resulted in nearly overlapping light-off curves, suggesting that moderate variations in ion flux density are not decisive for the catalytic outcome.

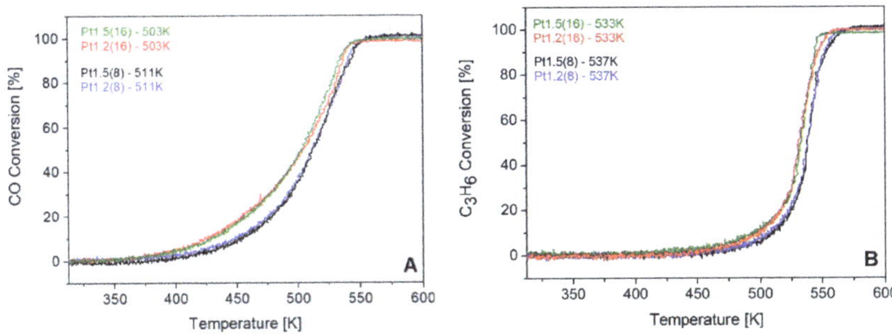

Figure 7. CO (**A**) and C_3H_6 (**B**) light-off curves over fresh Pt 1 wt.% $Ce_{0.5}Zr_{0.5}O_2$ catalyst compared with samples treated with an ion beam of 1.2 keV and 1.5 keV.

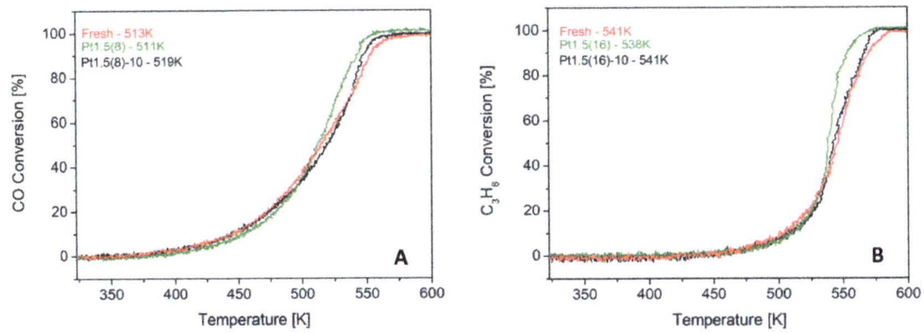

Figure 8. Comparison of light-off curves for samples treated with eight ion beam sputtering cycles at 1.5 keV and a current of 10 mA Pt1.5(8)-10 (**A**) and the sample treated under the same conditions with 20 mA (Pt1.5(8)) (**B**).

These experimental findings are consistent with the SRIM depth distribution profiles (Figure 9), which show that at 1.5 keV, atomic mixing extends throughout an entire 10 nm grain, whereas at 0.6 keV, the effect is largely confined to the outermost surface layers. Since catalytic activity is governed by near-surface phenomena, both energies were sufficient to induce the oxygen vacancy formation necessary for activity enhancement, explaining why small variations in beam energy within the tested range did not produce significant differences in light-off curves.

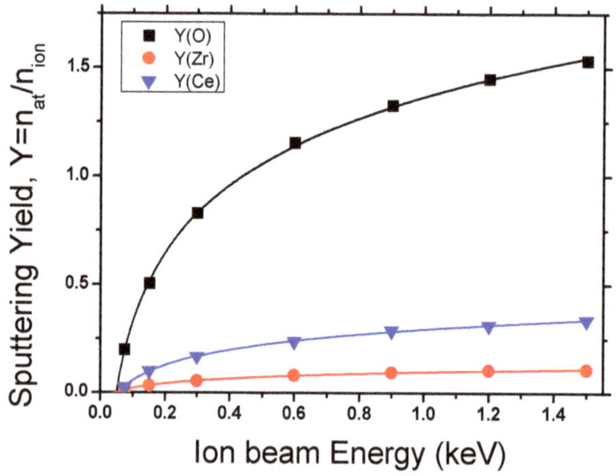

Figure 9. Correlation between SRIM simulations and experimental data, evidencing preferential oxygen sputtering and the stabilization of Ce^{3+}–V_O sites as the key factor behind the observed catalytic enhancement.

By contrast, the duration of each treatment emerged as a critical factor. When 16 treatments were applied with standard 15-min exposures (Pt1.5(16)), significant improvements were observed (CO T_{50} = 502 K, C_3H_6 T_{50} = 532 K). However, shortening the exposure to only 2 min per cycle (Pt1.5(16)-2) almost completely suppressed the beneficial effect, with light-off values (CO T_{50} = 514 K, C_3H_6 T_{50} = 546 K) close to the fresh catalyst (Figure 10). These findings confirm that a sufficient ion–surface interaction time is required to generate a meaningful concentration of oxygen vacancies and to stabilize Pt at the interface.

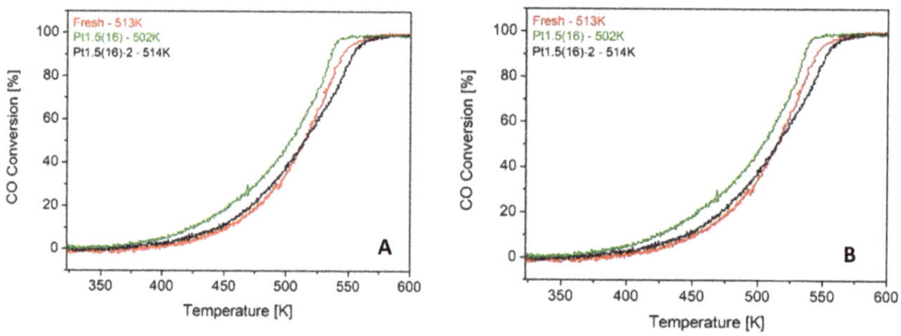

Figure 10. CO (**A**) and C3H6 (**B**) light-off curves over fresh Pt 1 wt.% $Ce_{0.5}Zr_{0.5}O_2$ catalyst compared with Pt1.5(16) and Pt1.5(16)-2, which had been exposed to the ion beam for 15 min and 2 min, respectively, during each treatment.

Taken together, these results establish a clear structure–activity correlation. Ion beam sputtering selectively removes oxygen atoms from the ceria lattice, generating oxygen vacancies and increasing the Ce^{3+} fraction. These defects act as anchoring sites for Pt, preventing migration and sintering, while simultaneously enhancing the reducibility and oxygen mobility of the support. The resulting Pt–Ce^{3+} interfacial synergy lowers the activation barrier for CO and hydrocarbon oxidation, leading to earlier light-off under lean-burn conditions. The fact that bulk crystallinity, surface area, and zirconia oxidation state remain unaffected highlights the surface-confined and tunable nature of this modification.

The simulations also provided quantitative estimates consistent with the experimental mass loss of treated samples, confirming that oxygen removal dominates the sputtering process under the tested conditions. The preferential sputtering of oxygen promotes the stabilization of Ce^{3+}–V_O sites, which act as anchoring points for Pt species and enhance Pt–ceria interfacial synergy. Together with the experimental evidence, the simulation results reinforce the conclusion that catalytic improvement arises from controlled, surface-confined oxygen depletion rather than bulk structural changes.

Moreover, the parameter study demonstrates that catalytic improvements are controlled by the extent of surface exposure—dictated by the number and duration of treatments—rather than by small variations in beam energy or current. This confirms that the decisive factor is the accumulation of sufficient surface defects rather than changes in penetration depth or ion flux density.

The present findings align with previous literature on [35], but extend this knowledge to a commercially relevant Pt/Ce–Zr system under realistic lean-burn exhaust conditions. By demonstrating reproducible and controllable reductions in light-off temperature, ion beam sputtering emerges as a promising post-synthesis strategy to fine-tune the surface chemistry of existing catalysts. Crucially, this approach provides a scalable means of addressing the persistent challenge of cold-start emissions in high-efficiency diesel and lean-burn engines, offering a pathway towards the development of next-generation emission-control systems.

3. Materials and Methods

3.1. Catalyst Preparation

A commercial $Pt/Ce_{0.5}Zr_{0.5}O_2$ catalyst (1 wt.% Pt, "Fresh"), commercially supplied by Rhodia (now Solvay, Brussels, Belgium), with detailed characterization in [36], was subjected to ion beam irradiation under controlled conditions. A 600 mg sample was placed in a 2 cm^2 holder and irradiated in a vacuum chamber using a Kaufman ion source (monoenergetic beam, 90° incidence). To ensure uniform exposure, the catalyst was periodically mixed between irradiation cycles to present fresh surface areas to the beam. The irradiation parameters—ion energy, beam current, exposure time, and number of treatment cycles—were systematically varied, as summarized in Table 1. The resulting samples were designated as PtX(Y), where X corresponds to the applied ion energy (keV) and Y to the number of irradiation treatments.

Specifically, the notation Pt1.5(16 × 4) indicates that four independent batches of approximately 100 mg each were irradiated under identical conditions (beam current 20 mA, ion energy 1.5 keV). Each batch underwent 16 consecutive irradiation runs of 15 min, ensuring reproducibility and providing sufficient treated material for subsequent analyses. Representative examples include Pt1.5(16 × 4), Pt1.5(8)-10, and Pt1.5(16)-2.

3.2. Catalytic Testing

Catalytic performance was assessed under conditions representative of lean-burn engine exhaust. Light-off experiments were conducted to determine the T_{50} values for CO and C_3H_6 oxidation. Tests were carried out in a fixed-bed quartz microreactor (inner diameter: 0.6 cm; length: 30 cm) loaded with 0.1 g of catalyst and housed in a tubular furnace. Reactor temperature was regulated using a Eurotherm (Ashburn, VA, USA) PID controller and monitored by a K-type thermocouple positioned at the center of the catalyst bed. The feed gas composition was maintained at a total flow rate corresponding to 16.7 $cm^3\ g^{-1}\ s^{-1}$, consisting of 10% O_2, 4.5% H_2O, 2000 ppm CO, 2000 ppm CH_4, 2000 ppm C_3H_6, and 200 ppm NO, with Ar as the balance gas. The reactor temperature was ramped from 313 to 773 K at a heating rate of 0.083 K s^{-1}. Product gases were continuously analyzed using a Pfeiffer OmniStar (Asslar, Germany) quadrupole mass spectrometer. To ensure reproducibility, three consecutive light-off cycles were performed; catalytic data presented herein correspond to the third cycle.

To complement the experimental characterization, simulations of N^+ ion bombardment were performed using SRIM-2013 (Stopping and Range of Ions in Matter—James F. Ziegler, Newtown, PA, USA). A $Ce_{0.7}Zr_{0.3}O_2$ lattice model was used to approximate the mixed oxide support composition. Ion energies of 0.6 and 1.5 keV were selected to reproduce the experimental conditions applied during ion beam sputtering, with normal incidence (90°) and a fluence consistent with the number of treatment cycles reported in Table 1. The simulations provided depth profiles of implanted ions and recoil distributions, as well as sputtering yields for O, Ce, and Zr atoms. These data were used to quantify the preferential removal of oxygen compared to the cations, to estimate the degree of atomic mixing within ~10 nm grains, and to rationalize the experimental findings of enhanced reducibility and increased Ce^{3+} concentration.

3.3. Material Characterization

The structural and physicochemical properties of the catalysts were investigated using a combination of ex situ techniques. X-ray diffraction (XRD) patterns were collected on a PANalytical (Malvern, UK) X'Pert Pro diffractometer equipped with Cu Kα radiation (40 kV, 40 mA), scanning over the 2θ range of 15–90° with a step size of 0.017°, with samples mounted on quartz zero-background holders. Textural properties, including specific surface area, pore volume, and pore size distribution, were determined by N_2 adsorption–desorption at 77 K using a Micromeritics (Norcross, GA, USA) ASAP 2010 instrument after sample degassing at 393 K under vacuum for 24 h. Platinum loading was quantified by energy-dispersive X-ray fluorescence (EDXRF) using a Rigaku (Tokyo, Japan) NEX QC+ QuantEZ analyzer. Temperature-programmed reduction (TPR) experiments were conducted following pretreatment in 3% O_2 (ambient temperature to 773 K, 5 K min^{-1}, 90 min isothermal hold), cooling under Ar, and subsequent exposure to 4000 ppm H_2 in Ar at a flow rate of 16.7 cm^3 g^{-1} s^{-1}. X-ray photoelectron spectroscopy (XPS) was performed on a Kratos Axis SUPRA instrument (Kratos Analytical Ltd., Manchester, UK) with Al Kα excitation (1486.69 eV, 12 mA, 15 kV, 700 × 300 µm analysis area). High-resolution spectra were acquired at a pass energy of 20 eV with 0.1 eV step size and 60 s acquisition time, employing an electron flood gun for charge compensation. Binding energies were calibrated to the C 1s peak at 284.8 eV, and data analysis was conducted using CasaXPS (v2.3.19PR1.0). Extended X-ray absorption fine structure (EXAFS) spectroscopy was performed at the Pt L_3-edge at the Diamond Light Source (UK), with first-shell fitting applied to determine Pt–O coordination environments.

4. Conclusions

This work demonstrates the effectiveness of ion beam sputtering as a post-synthesis surface engineering strategy for Pt/$Ce_{0.5}Zr_{0.5}O_2$ catalysts employed in automotive emission control. Comprehensive structural and spectroscopic analyses confirmed that low-energy ion irradiation selectively modified the near-surface region by generating oxygen vacancies and increasing the Ce^{3+} fraction, while preserving the bulk crystallinity and textural properties of the support. These surface modifications enhanced Pt–ceria interactions and improved redox behavior, as evidenced by shifts in TPR profiles and XPS analysis. Catalytic testing under lean-burn exhaust conditions revealed consistent reductions in light-off temperatures for CO and C_3H_6 oxidation, with the extent of improvement directly correlated to the cumulative ion exposure (number and duration of treatments). In contrast, variations in beam energy and current within the investigated ranges had negligible effects on performance, underscoring that surface activation is primarily governed by the degree of exposure rather than the specific irradiation parameters. Overall, the findings establish ion beam sputtering as a reproducible and controllable approach to fine-tune catalyst surfaces at the atomic scale. By directly linking surface defect engineering to measurable enhancements in low-temperature activity, this study highlights the potential of ion beam processing to address the persistent challenge of cold-start emissions in high-efficiency diesel and lean-burn engines. The strategy provides a precise, non-synthetic route to optimize existing commercial catalysts, offering new opportunities for the design of next-generation emission-control materials compliant with forthcoming Euro 7 regulations.

Author Contributions: R.O.: Methodology, formal analysis, investigation, data curation, visualization, M.M.: Methodology, formal analysis, resources, data curation, writing—original draft preparation, visualization, S.S.: formal analysis, data curation, visualization, N.A.: Conceptualization, Methodology, validation, resources, data curation, writing—original draft preparation, project administration, supervision. All authors have read and agreed to the published version of the manuscript.

Funding: This research received no external funding.

Data Availability Statement: The original contributions presented in this study article. Further inquiries can be directed to the corresponding author.

Acknowledgments: The authors gratefully acknowledge Mark Isaac for his valuable assistance in acquiring the EXAFS data.

Conflicts of Interest: The authors declare no conflicts of interest.

Abbreviations

The following abbreviations are used in this manuscript:

BET	Brunauer–Emmett–Teller (surface area measurement)
DPF	Diesel Particulate Filter
DOC	Diesel Oxidation Catalyst
EXAFS	Extended X-ray Absorption Fine Structure
HC	Hydrocarbons
LNT	Lean NOx Trap
OSC	Oxygen Storage Capacity
SCR	Selective Catalytic Reduction
SRIM	Stopping and Range of Ions in Matter
TPR	Temperature-Programmed Reduction
XPS	X-ray Photoelectron Spectroscopy
XRD	X-ray Diffraction
XRF	X-ray Fluorescence
V_O	Oxygen vacancy
PID	Proportional–Integral–Derivative (temperature controller)

References

1. ICCT. *A Technical Summary of Euro 6/VI Vehicle Emission Standards*; International Council for Clean Transportation: Washington, DC, USA, 2016.
2. Maddaloni, M.; Marchionni, M.; Abbá, A.; Mascia, M.; Tola, V.; Carpanese, M.P.; Bertanza, G.; Artioli, N. Exploring the Viability of Utilizing Treated Wastewater as a Sustainable Water Resource for Green Hydrogen Generation Using Solid Oxide Electrolysis Cells (SOECs). *Water* **2023**, *15*, 2569. [CrossRef]
3. La Corte, D.; Maddaloni, M.; Vahidzadeh, R.; Domini, M.; Bertanza, G.; Ansari, S.U.; Marchionni, M.; Tola, V.; Nancy, N.A. Recovered Ammonia as a Sustainable Energy Carrier: Innovations in Recovery, Combustion, and Fuel Cells. *Energies* **2025**, *18*, 508. [CrossRef]
4. Huang, S.; Shen, J.; Wu, Y.; Li, X.; Ma, Y.; Xie, Y.; Yu, C.; Zhang, Y.; Zhang, J. $Bi_2O_2CO_3$ Co-Catalyst Modification BiOBr Driving Efficient Photoreduction CO_2. *Colloids Surf. A Physicochem. Eng. Asp.* **2025**, *725*, 137731. [CrossRef]

5. Lietti, L.; Righini, L.; Castoldi, L.; Artioli, N.; Forzatti, P. Labeled ^{15}NO Study on N_2 and N_2O Formation over Pt-Ba/Al_2O_3 NSR Catalysts. *Top. Catal.* **2013**, *56*, 7–13. [CrossRef]
6. Maddaloni, M.; Centeno-Pedrazo, A.; Avanzi, S.; Mazumdar, N.J.; Manyar, H.; Artioli, N. Novel Ionic Liquid Synthesis of Bimetallic Fe–Ru Catalysts for the Direct Hydrogenation of CO_2 to Short Chain Hydrocarbons. *Catalysts* **2023**, *13*, 1499. [CrossRef]
7. Di Monte, R.; Kašpar, J. Heterogeneous Environmental Catalysis—A Gentle Art: CeO_2-ZrO_2 Mixed Oxides as a Case History. *Catal. Today* **2005**, *100*, 27–35. [CrossRef]
8. Montini, T.; Melchionna, M.; Monai, M.; Fornasiero, P. Fundamentals and Catalytic Applications of CeO_2-Based Materials. *Chem. Rev.* **2016**, *116*, 5987–6041. [CrossRef]
9. O'Donnell, R.; Ralphs, K.; Grolleau, M.; Manyar, H.; Artioli, N. Doping Manganese Oxides with Ceria and Ceria Zirconia Using a One-Pot Sol–Gel Method for Low Temperature Diesel Oxidation Catalysts. *Top. Catal.* **2020**, *63*, 351–362. [CrossRef]
10. Karre, A.V.; Garlapalli, R.K.; Jena, A.; Tripathi, N. State of the Art Developments in Oxidation Performance and Deactivation of Diesel Oxidation Catalyst (DOC). *Catal. Commun.* **2023**, *179*, 106682. [CrossRef]
11. Lafyatis, D.S.; Ansell, G.P.; Bennett, S.C.; Frost, J.C.; Millington, P.J.; Rajaram, R.R.; Walker, A.P.; Ballinger, T.H. Ambient Temperature Light-off for Automobile Emission Control. *Appl. Catal. B Environ.* **1998**, *18*, 123–135. [CrossRef]
12. Coise Duprat, F. Light-off Curve of Catalytic Reaction and Kinetics. *Chem. Eng. Sci.* **2002**, *57*, 901–911. [CrossRef]
13. Li, T.; Zhao, P.; He, H.; Wang, C.; Zhang, H.; Chen, Z.; Chen, H. Dual-Fuel Dual-Direct Injection: An Efficient and Clean Combustion Technology for Diesel Engines. *J. Energy Inst.* **2025**, *119*, 102006. [CrossRef]
14. Dardiotis, C.; Martini, G.; Marotta, A.; Manfredi, U. Low-Temperature Cold-Start Gaseous Emissions of Late Technology Passenger Cars. *Appl. Energy* **2013**, *111*, 468–478. [CrossRef]
15. Ye, S.; Yap, Y.H.; Kolaczkowski, S.T.; Robinson, K.; Lukyanov, D. Catalyst "light-off" Experiments on a Diesel Oxidation Catalyst Connected to a Diesel Engine-Methodology and Techniques. *Chem. Eng. Res. Des.* **2012**, *90*, 834–845. [CrossRef]
16. Solt, H.; Maddaloni, M.; Bazin, P.; Aureau, D.; Etcheberry, A.; Busardo, D.; Rousseau, S.; Blanchard, G.; Moral, N.; Bruma, A.; et al. Surface Modification of Nanocatalysts via Ion Beam Techniques for Enhanced Activity. *Appl. Catal. A Gen.* **2025**, *707*, 120536. [CrossRef]
17. Magudapathy, P.; Srivastava, S.K.; Gangopadhyay, P.; Amirthapandian, S.; Saravanan, K.; Das, A.; Panigrahi, B.K. Alloying of Metal Nanoparticles by Ion-Beam Induced Sputtering. *Chem. Phys. Lett.* **2017**, *667*, 38–44. [CrossRef]
18. Fu, K.; Zeng, L.; Liu, J.; Liu, M.; Li, S.; Guo, W.; Gao, Y.; Pan, M. Magnetron Sputtering a High-Performance Catalyst for Ultra-Low-Pt Loading PEMFCs. *J. Alloys Compd.* **2020**, *815*, 152374. [CrossRef]
19. Wang, G.D.; Kong, D.D.; Pan, Y.H.; Pan, H.B.; Zhu, J.F. Low Energy Ar-Ion Bombardment Effects on the CeO_2 Surface. *Appl. Surf. Sci.* **2012**, *258*, 2057–2061. [CrossRef]
20. Liu, Y.; Li, F.; Tian, H.; Wang, G.; Wang, X. Influence of Ion Beam Surface Treatment on the Emission Performance of Photocathodes. *Nanoscale Adv.* **2022**, *4*, 3517–3523. [CrossRef]
21. Clark, A.H.; Marchbank, H.R.; Thompsett, D.; Fisher, J.M.; Longo, A.; Beyer, K.A.; Hyde, T.I.; Sankar, G. On the Effect of Metal Loading on the Reducibility and Redox Chemistry of Ceria Supported Pd Catalysts. *Phys. Chem. Chem. Phys.* **2022**, *24*, 2387–2395. [CrossRef]
22. Kimata, T.; Kakitani, K.; Yamamoto, S.; Shimoyama, I.; Matsumura, D.; Iwase, A.; Mao, W.; Kobayashi, T.; Yamaki, T.; Terai, T. Activity Enhancement of Platinum Oxygen-Reduction Electrocatalysts Using Ion-Beam Induced Defects. *Phys. Rev. Mater.* **2022**, *6*, 035801. [CrossRef]
23. Wan, W.; Geiger, J.; Berdunov, N.; Lopez Luna, M.; Chee, S.W.; Daelman, N.; López, N.; Shaikhutdinov, S.; Roldan Cuenya, B. Highly Stable and Reactive Platinum Single Atoms on Oxygen Plasma-Functionalized CeO2 Surfaces: Nanostructuring and Peroxo Effects. *Angew. Chem. Int. Ed.* **2022**, *61*, e202112640. [CrossRef]
24. Zlobin, V.N.; Bannikov, M.G.; Vasilev, I.P.; Cherkasov, J.A.; Gawrilenko, P.N. Potential of Use of Ion Implantation as a Means of Catalyst Manufacturing. *Proc. Inst. Mech. Eng. Part D J. Automob. Eng.* **2002**, *216*, 385–390. [CrossRef]
25. Qu, J.; Liu, W.; Liu, R.; He, J.; Liu, D.; Feng, Z.; Feng, Z.; Li, R.; Li, C. Evolution of Oxygen Vacancies in Cerium Dioxide at Atomic Scale under CO2 Reduction. *Chem. Catal.* **2023**, *3*, 100759. [CrossRef]
26. Wolf, G.K. Chemical and Catalytic Effects of Ion Implantation. *Radiat. Eff.* **2006**, *65*, 107–116. [CrossRef]
27. Machida, M.; Taniguchi, H.; Kijimaa, T.; Nakatanib, J. Methane Combustion Activity of Alumina Supported Pt, Pd, and Rh Modified by High-Energy Ion Beam Irradiation. *J. Mater. Chem.* **1998**, *8*, 781–785. [CrossRef]
28. Shah, P.M.; Burnett, J.W.H.; Morgan, D.J.; Davies, T.E.; Taylor, S.H. Ceria–Zirconia Mixed Metal Oxides Prepared via Mechanochemical Grinding of Carbonates for the Total Oxidation of Propane and Naphthalene. *Catalysts* **2019**, *9*, 475. [CrossRef]
29. Parthasarathi, B.; Priolkar, R.K.; Gayen, A.; Sarode, P.R.; Hegde, M.S.; Emura, S.; Kumashiro, R.; Jayaram, V.; Subbanna, G.N. Ionic Dispersion of Pt over CeO_2 by the Combustion Method Structural Investigation by XRD, TEM, XPS, and EXAFS. *Chem. Mater.* **2003**, *15*, 2049–2060. [CrossRef]

30. Liotta, L.F.; Longo, A.; Macaluso, A.; Martorana, A.; Pantaleo, G.; Venezia, A.M.; Deganello, G. Influence of the SMSI Effect on the Catalytic Activity of a Pt(1%)/$Ce_{0.6}Zr_{0.4}O_2$ Catalyst: SAXS, XRD, XPS and TPR Investigations. *Appl. Catal. B* **2004**, *48*, 133–149. [CrossRef]
31. Ji, Y.; Xu, D.; Bai, S.; Graham, U.; Crocker, M.; Chen, B.; Shi, C.; Harris, D.; Scapens, D.; Darab, J. Pt- and Pd-Promoted CeO_2-ZrO_2 for Passive NO_x Adsorber Applications. *Ind. Eng. Chem. Res.* **2017**, *56*, 111–125. [CrossRef]
32. Romeo, M.; Bak, K.; El Fallah, J.; Le Normand, F.; Hilaire, L. XPS Study of the Reduction of Cerium Dioxide. *Surf. Interface Anal.* **1993**, *20*, 508–512. [CrossRef]
33. Osazuwa, O.U.; Ng, K.H. The Roles of Oxygen Mobility and Oxygen Vacancy in Metallic Catalysts-Prompted Dry Reforming of Methane: A Review. *Renew. Energy* **2026**, *256*, 124074. [CrossRef]
34. Wisniewska, J.; Ziolek, M.; Artioli, N.; Daturi, M. The Effect of Niobium and Tantalum on Physicochemical and Catalytic Properties of Silver and Platinum Catalysts Based on MCF Mesoporous Cellular Foams. *J. Catal.* **2016**, *336*, 58–74. [CrossRef]
35. Kalia, S.; Salim, A.; Bala, R.; Singh, R.K.; Singhal, R.; Dhiman, R. Low Energy Ag Ion Irradiation-Induced Surface Modification Studies of Au-Graphene Oxide Nanocomposite Thin Films. *Thin Solid Films* **2023**, *772*, 139802. [CrossRef]
36. Colo, G.; Pijolat, M.; Valdivieso, F.; Vidal, H.; Kas-Par, J.; Finocchio, E.; Daturi, M.; Binet, C.; Lavalley, J.C.; Bakerd, R.T.; et al. Surface and Structural Characterization of Ce x $Zr_{1\text{-}x}O_2$ CEZIRENCAT Mixed Oxides as Potential Three-Way Catalyst Promoters. *J. Chem. Soc. Faraday Trans.* **1998**, *94*, 3717–3726. [CrossRef]

catalysts

MDPI

Article

Magnetic Cross-Linked Enzyme Aggregates of Glycoside Hydrolase: An Efficient and Stable Biocatalyst for Icaritin

Yuxuan Zhao [1], Wei Zhang [1], Ye Li [2,*], Wenting Fei [2] and Hao Liang [1,*]

[1] State Key Laboratory of Chemical Resource Engineering, Beijing University of Chemical Technology, Beijing 100029, China
[2] College of Bioengineering, Beijing Polytechnic University, Beijing 100176, China; fwt911021@163.com
* Correspondence: liyeyashi@163.com (Y.L.); lianghao@mail.buct.edu.cn (H.L.)

Abstract: With the booming development of natural anticancer agents, icaritin has been widely used in clinical liver cancer treatment due to the smaller number of side effects. Among them, enzymes as catalysts for producing icaritin have attracted considerable attention. Industrial production remains in its infancy due to the poor reusability of free enzymes, despite enzymes possessing favorable efficiency and green catalytic effects. The present study investigated two immobilization methods, including cross-linked enzyme aggregates (CLEAs) and magnetic cross-linked enzyme aggregates (MCLEAs). The optimal temperature and pH for the two hydrolases were determined, followed by characterization using VSM, XRD, FT-IR, and TEM techniques. The experiments have demonstrated that MCLEAs eradicate the need for ultrafiltration; MCLEAs are beneficial for enhancing fixation efficiency. Additionally, MCLEAs exhibited significantly higher catalytic activity, which raised the catalytic activity by approximately 30% compared with CLEAs. Moreover, after 10 consecutive reuse cycles, the catalytic activity of MCLEAs remained above 70%, maintaining a conversion rate of epimedin C at 61.59%. To summarize, MCLEAs offer an efficient strategy for enzyme immobilization. MCLEAs not only significantly enhanced both the enzyme's ability to catalyze and resist but also eliminated the necessity for ultrafiltration as well as enabled rapid product separation.

Keywords: glycosidase; immobilized enzyme; magnetic nanoparticles

1. Introduction

Globally, hepatocellular carcinoma is ranked as the sixth most common cancer and holds the fourth position in terms of causing mortality related to cancer [1]. The predominant strategy for most cancer management is chemotherapy, which relies on the potent cytotoxicity of chemotherapeutic agents against malignant cells. However, the inherent limitations and significant side effects of conventional chemotherapy drugs have consistently been demonstrated [2]. In an endeavor to mitigate the side effects and adverse reactions associated with chemotherapeutic medications, researchers are currently exploring the potential of harnessing natural anticancer compounds derived from diverse botanical origins and organisms. Their primary objective is to attain optimal efficacy while minimizing toxicity levels [3]. Epimedium, a well-known herbal medicine in China, is frequently used as a tonic and treatment for rheumatism [4]. Icaritin (ICT), a primary flavone compound with an 8-prenyl group, has been identified as a significant component found in medicinal plants belonging to the Epimedium genus. Its effectiveness in cancer treatments has been demonstrated through mechanisms associated with apoptosis [5]. In the year 2022, a molecular-level immunomodulatory drug in the form of an icaritin soft

 https://doi.org/10.3390/catal15111034

capsule was introduced to the market, offering a viable therapeutic option for individuals suffering from hepatocellular carcinoma and unfavorable prognosis. Icaritin soft capsules significantly improved the standard of life in individuals diagnosed with hepatocellular carcinoma [6]. Icaritin showed satisfactory clinical performance and possessed great global market prospects; thus, icaritin has attracted widespread attention from both academia and medicine.

Due to the significantly limited presence of icaritin (less than 1%) in epimedium, the direct extraction method is considerably tough to prepare icaritin on a large scale. In addition, the synthesis of icaritin through chemical means is inefficient because of the costly catalysts, harsh conditions, intricate procedure, and complex byproducts [7,8]. Natural enzymes as green catalysts are a promising method for producing icaritin. At present, the enzymes have been used to prepare icaritin, mainly including glucosidase, rhamnosidase, snail enzyme, cellulase, pectinase, and naringinase [9–11].

As the majority of published research used icariin as a substrate, while the extraction method of icariin is complicated, and the yield is not high. In contrast to icariin, epimedin C is more abundant in epimedium, so the cost of preparing icaritin can be greatly decreased by using epimedin C as a substrate [12,13]. Although epimedin C is present at a relatively high content level, its hydrolysis is hindered by the two rhamnose molecules at the C3 position, making it a less explored target. Recently, Xie et al. were the first to report the use of β-glucosidase and α-rhamnosidase to achieve the stepwise hydrolysis of epimedin C, ultimately yielding icaritin [9]. β-glucosidase (G4, MT779019) causes the hydrolysis of epimedin C by clearing the glycosidic bond at the C7 position. Additionally, α-rhamnosidase (R2, MT779021) possesses the ability to remove rhamnose moieties at the C3 position. The icaritin production from epimedin C uses R2 and G4 by a two-step catalysis. However, the reusability and operational stability of free enzymes were relatively restricted, which limited their potential industrial application [14]. Therefore, it is imminent to find a suitable approach to immobilization [15].

The immobilization of enzymes has gained significant attention because of its benefits in overcoming the demerits of free enzymes [16]. The past few decades have witnessed a growing body of investigation into the immobilization of enzyme methods, for example, carrier-bound immobilization methods including entrapment, adsorption, and covalence, which have several demerits [17]. In contrast, the benefits of the carrier-free immobilization approach were more obvious. Particularly, cross-linked enzyme aggregates (CLEAs) technology had arose as a viable option due to its reusability, ease of preparation, and capability to be detached from reactions in industrial applications [18]. Thus, it offers significant technical and economic benefits.

Application of CLEAs is constrained by certain limitations, which hinder its further implementation. One of the reasons is that the enhanced size (clumping phenomenon) of CLEAs clusters occurs as a result of their separation from a mixture of reactions through centrifugation or filtration. This phenomenon may potentially impair the mass transfer efficiency of enzymes, and such an effect is more prominent when enzymes act on macromolecular substrates [19]. The mechanical instability of CLEAs poses another concern, as it leads to the loss of enzyme activity during filtration and centrifugation [20].

To overcome this, iron oxide magnetic nanoparticles (MNPs) have attracted much interest. The immobilization of CLEAs using MNPs has been extensively employed in numerous studies to prepare magnetic cross-linked enzyme aggregates (MCLEAs) [21].

In addition to the added advantage of facile separation and recovery from reaction media, MNPs lower operating costs compared to traditional methods like filtration and centrifugation [22]. The aforementioned nanomaterials exhibit distinctive characteristics of biocompatibility and non-toxicity. MNPs are superior supports for enzyme immobilization, offering a substantial area on the surface for potentially elevated enzyme loading [23]. The choice of a cross-linking agent plays a vital role in the formation of enzyme aggregates [24]. Glutaraldehyde, known for its remarkable versatility, high reactivity, and stability, is widely used as a cross-linking agent to immobilize various types of enzymes [25]. Through covalent bonds formed by cross-linking, it has the potential to decrease thermal oscillation and limit conformational adaptability, thereby mitigating the denaturation and unfolding of enzymes [26]. To summarize, these magnetic cross-linked enzyme aggregates (MCLEAs) have demonstrated enhanced stability against pH and temperature variations, reduced inhibition effects, and high activity levels [27]. This makes them an indispensable strategy in improving enzymatic processes.

This study investigated various immobilization techniques for glycosidases, including the synthesis of cross-linked enzymatic aggregates (CLEAs) and magnetic cross-linked enzyme aggregates (MCLEAs). CLEAs were prepared by ammonium sulfate $(NH_4)_2SO_4$ precipitation, glutaraldehyde cross-linking, and centrifugation. For MCLEAs, magnetic nanoparticles (MNPs) were used, and the aggregates were collected by magnetic separation after $(NH_4)_2SO_4$ precipitation and glutaraldehyde cross-linking (without ultrafiltration, Scheme 1).

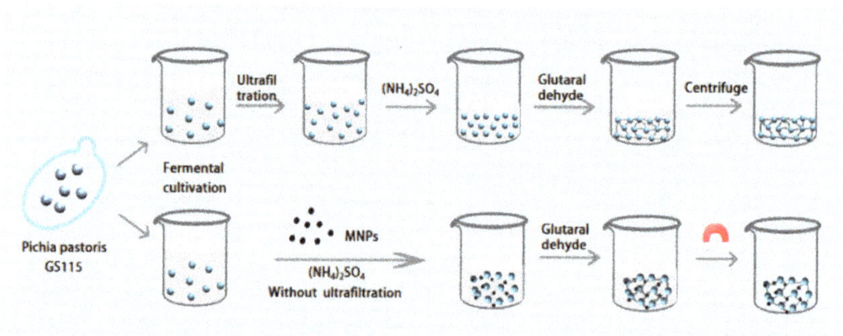

Scheme 1. Flow chart of the two preparation routes.

The optimization of preparation conditions for immobilizing R2 and G4 was conducted by factors such as precipitation, cross-linking conditions, and preparation pH on activity recovery. Furthermore, a systematic comparison was made between the enzyme properties of CLEAs and MCLEAs.

2. Results

2.1. Optimization and Characterization of CLEAs

The addition of salts to aqueous protein solutions resulted in the precipitation of physical aggregates, which remained together due to non-covalent bonding without causing any disruption to their tertiary structure [28]. The precipitation of enzymes using ammonium sulfate is a widely employed purification method. Thus, the precipitant chosen for the subsequent experiment was saturated ammonium sulfate. The correct dosage of precipitant was determined by investigating various volume ratios of ammonium sulfate. As shown in Figure 1A, the ratio gradient of the fermentation broth to ammonium sulfate was from 1:0.25 to 1:3. The highest enzyme activity recovery was achieved when the ratio reached 1:2. As ammonium sulfate continued to increase, the enzyme activity gradually decreased. A reduced dosage will result in ineffective enzyme precipitation, while an increased dosage may lead to enzyme denaturation due to the removal of the hydration layer, resulting in the loss of enzymatic activity [29].

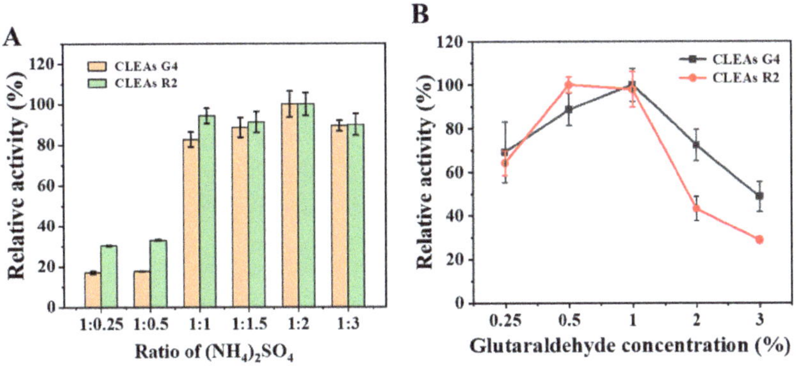

Figure 1. Optimization of CLEAs preparation. Impact of different concentrations of $(NH_4)_2SO_4$ (**A**) and glutaraldehyde (**B**) on enzyme.

Glutaraldehyde is a powerful cross-linking agent due to its cost-effectiveness, ready availability in large quantities for commercial purposes, and status as one of the most popular reagents in biocatalytic design [30]. When utilized at various concentrations (0.25–2%) to form CLEAs, the activity recovery of CLEAs increased most substantially as the GA concentration was elevated to 1% (Figure 1B). Nevertheless, with further increases in the GA concentration, we observed a sharp decline in activity recovery, which may be owing to the infiltration of GA into the protein core, resulting in an adverse interaction with crucial catalytic amino acids [31,32]. The other factors contributing to the decrease in enzymatic activity at elevated GA concentrations may be the diminishment of enzyme flexibility or more compact CLEAs, which result in the limited diffusion of coarse substrates to the catalytic site [32,33].

The CLEAs can form either regular spherical aggregates (type 1) or asymmetric cross-linked aggregates (type 2) [34]. To investigate the structure of the CLEAs, SEM was used to capture micrographs. As depicted in Figure S1A, CLEAs G4 exhibited a relatively uniform structure comprising numerous small globular particles. This spherical morphology offers an increased specific surface area and a higher catalytic site per unit area, thereby facilitating their application as biocatalysts [10]. Conversely, CLEAs R2 (Figure S1B) displayed a less spherical morphology where proteins interlinked to create porous networks resembling type II structures. Furthermore, both types of CLEAs exhibited highly porous network structures, which enhance the connection between substrates and the active sites of enzymes for improved efficiency [35].

To further investigate the interaction between free enzymes and CLEAs, the secondary structure of CLEAs was analyzed using FT-IR spectroscopy, which was a crucial technique for monitoring the reconfiguration of protein conformational alterations in protein structures [16]. As shown in Figure S2, the characteristic peaks observed at 1510 cm^{-1} and 1560 cm^{-1} are attributed to the protein amide II band, which results from the coupled vibrations of C=N stretching and N–H bending [36]. The absorption at 1150 cm^{-1} is associated with C–O stretching vibrations originating from amino acid residues [37]. Additionally, the broad spectral feature in the 3400–3300 cm^{-1} region corresponds to O–H stretching vibrations, likely introduced by glutaraldehyde-mediated cross-linking [38]. Together, these spectral signatures provide clear evidence for the successful immobilization of glycosidases onto the magnetic nanoparticle support.

2.2. Optimization and Characterization of MCLEAs

The requirement for centrifugation to recover the immobilized enzyme remains one of the drawbacks associated with conventional CLEAs [39]. To solve this issue, we introduced magnetic Fe_3O_4 nanoparticles as cores, which were co-precipitated with the aggregated enzyme and then cross-linked, thus achieving the results of recycling with a magnet [40]. The magnetic cross-linked enzyme aggregate comprises an enzyme aggregate and a surface-functionalized magnetic particle, which are cross-linked together by glutaraldehyde [28]. The preparation method is to mix the enzyme solution and the surface-functionalized magnetic particles evenly, then add ammonium sulfate to make the enzyme form an enzyme aggregate, and finally add glutaraldehyde, serving as a cross-linking agent to cross-link the enzyme aggregate and the surface-functionalized magnetic particles to form the magnetic cross-linked enzyme aggregate [41]. In this experiment, we used 4 mg of carboxylated ferric oxide with 0.2U of R2 and G4 to prepare magnetic cross-linked enzyme aggregates.

First, we optimized the preparation conditions. Compared to CLEAs, MNPs offer a multitude of surface regions for enhanced enzyme loading, thereby obviating the need for ultrafiltration. MCLEAs were prepared by initially physically aggregating enzymes using precipitants and subsequently establishing chemical bonds between them through cross-linking agents. The optimization of the ammonium sulfate concentration is also crucial, as it can induce varying degrees of enzyme aggregation during precipitation. As shown in Figure 2A, when the volume fraction of the precipitant increased, the precipitated protein increased, and the enzyme activity recovery gradually increased. However, after reaching a certain ratio (1:3), even though the content of ammonium sulfate exhibited a continuous increase, the recovery of enzyme activity did not increase but began to decrease. High concentrations of ammonium sulfate enriched the solution with hydroxyl groups, which adversely affected the protein's tertiary structure and resulted in the reduced catalytic activity of glycosidases [42,43].

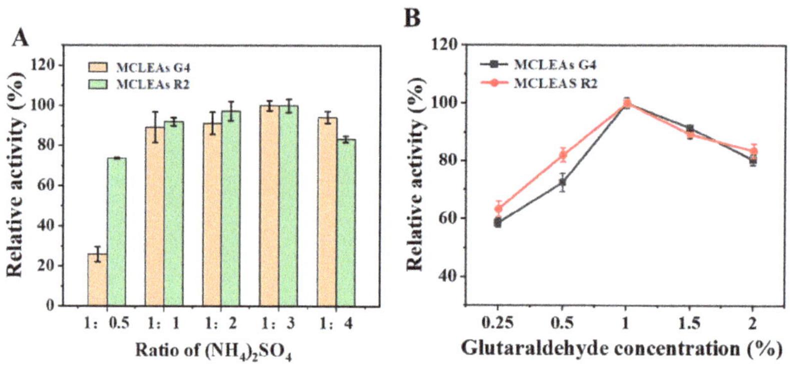

Figure 2. Optimization of MCLEAs preparation. Impact of different concentrations of $(NH_4)_2SO_4$ (**A**) and glutaraldehyde (**B**) on enzyme.

The concentration of GA is an important factor affecting the recovery of MCLEAs activity as well. At lower cross-linker concentrations, inadequate cross-linking occurs, releasing the enzyme into the reaction medium as a consequence of unstable CLEAs and MCLEAs [41]. The relative activity of MCLEAs was the highest at a final glutaraldehyde concentration of 1% (Figure 2B). However, when the glutaraldehyde concentration continued to improve, the relative activity of MCLEAs began to decrease. The reason for this is that elevated concentrations of glutaraldehyde can induce conformational changes in the active sites of enzymes [44]. Furthermore, in high concentrations of glutaraldehyde, the increased rigidity of enzymes lead to the loss of flexibility of the enzymes; therefore, the catalytic activity decreases [45].

The formation of MCLEAs was verified by performing TEM on carboxyl-functionalized Fe_3O_4 nanoparticles and MCLEAs. Panel A shows the magnetic nanoparticles, panel B shows the G4 on magnetic nanoparticles (MCLEAs G4), and the TEM of MCLEAs R2 is shown in the Figure S3. The electron beam in TEM analysis passes through a thin sample, and the contrast between denser and less dense regions is formed by the variation in electron density of the elements present in the sample, resulting in darker contrasts for higher electron densities [46]. As shown in Figure S3A, Fe_3O_4 nanoparticles are granular, uniformly dispersed, smooth, clearly delineated, and bright. In contrast, as shown in Figure S3B, after MCLEAs G4 formation, the contours were not clear, and overall transmittance decreased due to the encapsulation of surface proteins, proving that proteins were successfully attached to the MNPs surface.

The FTIR analysis (Figure S4) confirmed the presence of MNPs within the structure of MCLEAs, but the presence of proteins gives identifiable differences in their characteristic peaks. The absorption peaks observed at 570 cm^{-1} in the FT-IR spectrum of MCLEAs correspond to the stretching vibrations of Fe–O bonds, providing evidence for MNP incorporation into MCLEAs [47,48]. The confirmation of the enzyme's presence was established through the characteristic peak of the N-H bond at 3145 cm^{-1} and the characteristic peak of the C-O bond at 1082 cm^{-1} [49,50]. There are characteristic peaks at 1395 cm^{-1} and 1554 cm^{-1}, which are mainly formed by the amide bond after enzyme cross-linking [21,46]. Thus, the successful immobilization of enzymes onto magnetic nanoparticles can be confirmed.

The X-ray diffraction (XRD) technique was commonly employed to investigate the crystalline arrangement of magnetic nanoparticles, both prior to and subsequent to the immobilization process [51]. As depicted in Figure S5, the powder XRD patterns of the synthesized samples exhibited a close match in terms of the peak positions and relative intensities with the reference magnetite Fe_3O_4 pattern (JCPDS card 19-0629) [52,53]. These findings indicate that the cross-linking steps did not induce any structural alterations in MNPs.

The nanoparticles used in this study have the added advantage of easy separation. The magnetic behavior was assessed through VSM analysis (Figure S6). Based on the findings, it was observed that MNPs exhibit a saturation magnetization value of 62.88 emu/g, indicating their super magnetic property [48]. Additionally, the saturation magnetization values of MCLEAs G4 and MCLEAs R2 were measured to be 41.54 emu/g and 33.45 emu/g, respectively. The decline in the saturation magnetization value of MNPs indicates successful enzyme loading on MNPs. Despite the decline in the magnetic saturation value following immobilization, the prepared MCLEAs are quickly attracted towards the applied magnet.

2.3. *Assays of the Optimal pH and Temperature*

After determining the optimal cross-linking conditions, we investigated the optimal pH and temperature changes in MCLEAs and CLEAs. The activity of MCLEAs and CLEAs was investigated across a broad temperature range (30–80 °C) and pH range (3.0–7.0). Relative activity was achieved by establishing the maximum activity at 100%.

As shown in Figure 3A, G4 demonstrated its highest level of activity when the pH value was set at 4.0, and the optimum pH value remained unchanged before and after fixation. However, at pH 5.0 to 7.0, the relative activity of CLEAs G4 was above 20%, while the catalytic activity of the free G4 was significantly reduced, indicating that immobilized enzymes had a wider pH adaptability. Meanwhile, Figure 3B shows that the activity of R2 was the highest at pH 4.5, and the optimal pH did not change before and after fixation; however, the activity of immobilized enzymes was higher than free enzymes across various pH conditions. The alteration of the pH in the enzyme solution may induce subunit dissociation of the multimeric enzyme, thereby exerting a significant impact on its enzymatic activity [54]. At extreme pH, CLEAs can maintain greater activity than free enzymes because cross-linking enhanced the stability of enzyme structures and protected enzyme subunits from dissociation [55]. Next, we compared the change in the optimal temperature of the two enzymes before and after immobilization. As shown in Figure 3C, at a temperature of 55 °C, CLEAs R2 exhibited the greatest relative activity as the free enzyme. CLEAs R2 retained above 30% of its relative enzyme activity at various temperatures from 65 °C to 80 °C, indicating that CLEAs R2 has good thermal adaptability. For CLEAs G4, the optimum temperature was elevated to 10 °C compared to that of free G4, and its catalytic activity was still above 50% even at 75 °C. After immobilization, the enzyme exhibits enhanced rigidity and reduced molecular mobility, which is crucial for protecting the enzyme structure under harsh acidic and alkaline conditions [56]. In addition, the application of elevated temperatures can enhance the solubility of the substrate, which helps to accelerate the hydrolysis reaction and reduce the production cost [57].

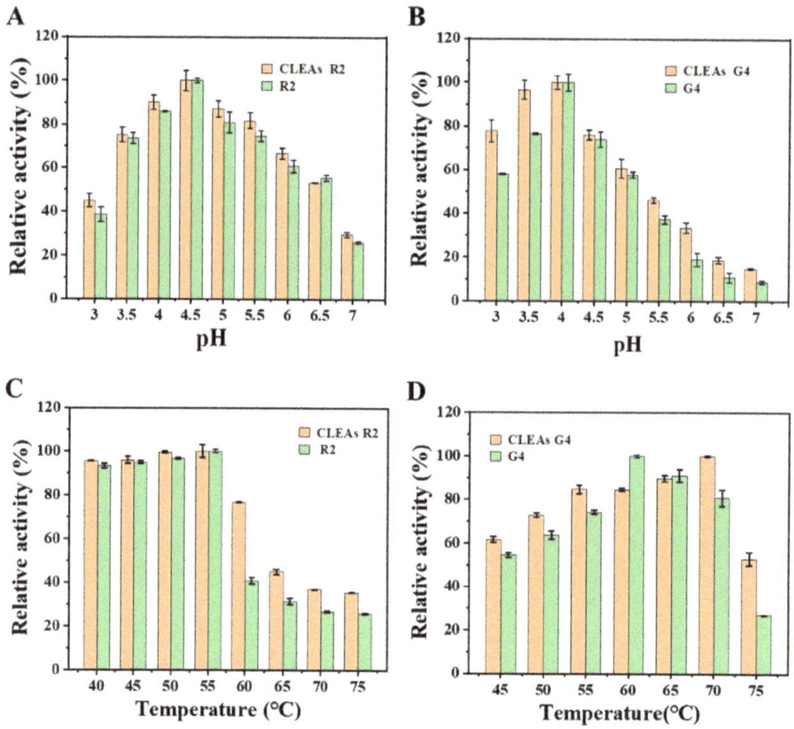

Figure 3. Effect of pH (**A**,**B**) and temperature (**C**,**D**) on free enzymes and CLEAs. The relative activity was expressed as the percentage of control activity.

In general, although the enzymatic properties of CLEAs are better than those of free enzymes, they do not perform particularly well. In contrast, MCLEAs has a better performance. The enzyme activity of MCLEAs was significantly improved under the extreme conditions of pH 3.0 and 7.0, as shown in Figure 4. In terms of the optimal temperature, MCLEAs also have a relatively good performance. This may be due to covalent binding to carboxylated ferric oxide, which may limit conformational changes and thus lead to higher relative activity over a wide pH range [40,58].The optimum temperature of R2 was raised from 55 °C to 60 °C. Under an extremely high temperature of 75 °C, the activity of free G4 was less than 30%, while that of MCLEAs G4 was nearly 60%, indicating that the enzyme activity was significantly increased. The observed enhancement may be attributed to the presence of nanoparticle pores that trap water molecules, thereby preserving their structure and improving enzyme stability [55].

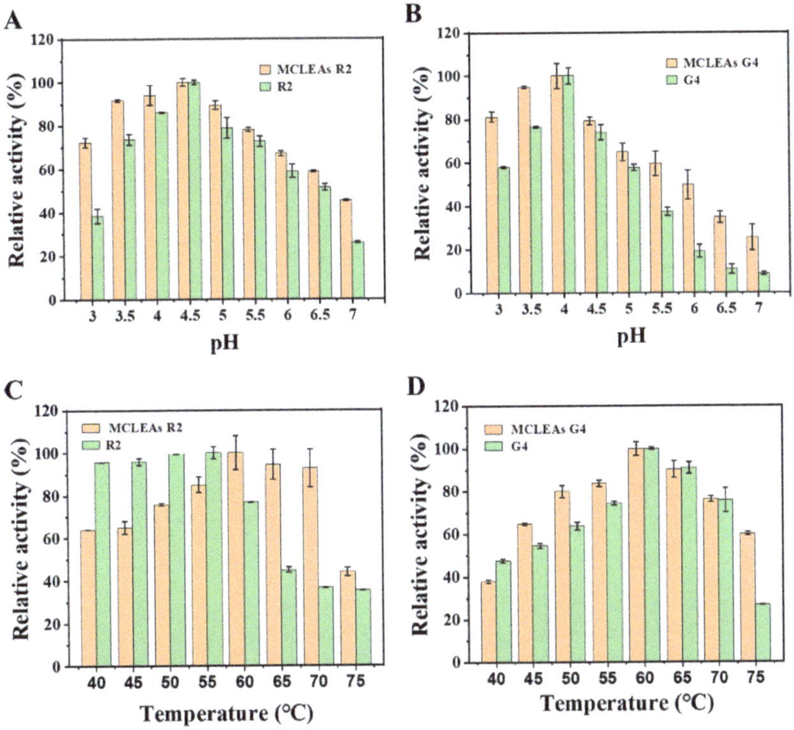

Figure 4. Effect of pH (**A**,**B**) and temperature (**C**,**D**) on MCLEAs and free enzymes. The relative activity was expressed as the percentage of control activity.

2.4. Stability and Reusability

Inactivation of enzymes under harsh conditions posed a crucial limitation to their widespread implementation in the industry. We conducted an investigation into the stability of cross-linked enzyme aggregates (CLEAs), magnetic cross-linked enzyme aggregates (MCLEAs), and free enzymes under varying pH conditions.

The residual activity of MCLEAs and CLEAs remained stable and did not exhibit a significant decrease across the tested pH range (Figure 5A,B) following a 2 h incubation. The incubation at a high temperature of 65 °C for 30–90 min, as depicted in Figure 5C,D, immobilized enzymes exhibited a significantly greater level of activity compared to free enzymes; however, MCLEAs demonstrated superior stability when compared to CLEAs. The significant residual activity in MCLEAs could potentially be ascribed to the sufficient cross-linking of the enzyme with MNPs [59]. The process of cross-linking increased the rigidity of the enzymes, resulting in a prominent improvement in protein's tolerance to its surrounding environment. The reusability of biocatalysts is a crucial element deciding industrial applicability, serving as a significant indicator of the effectiveness of immobilized enzymes.

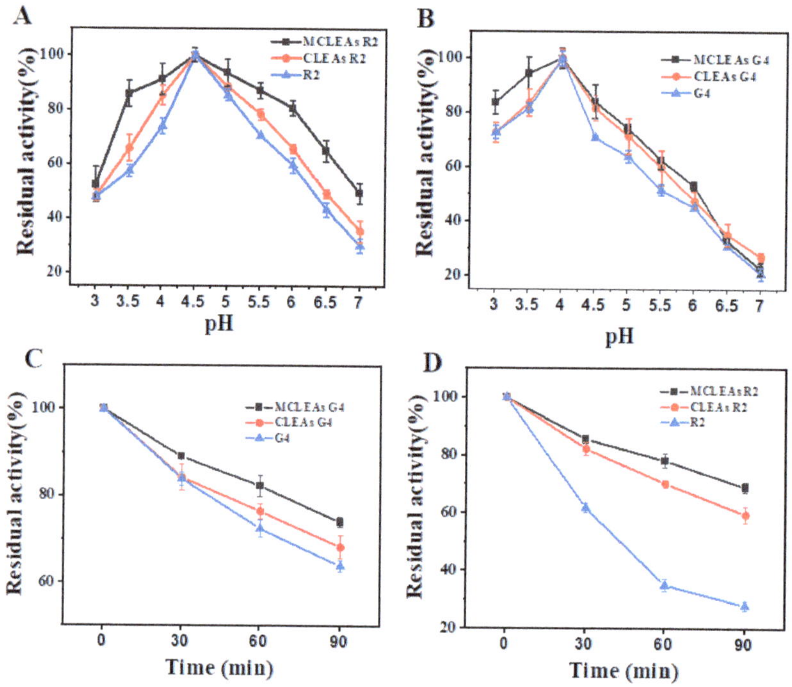

Figure 5. Evaluation of enzyme stability at different pH and high-temperature conditions. (**A**) Residual activity of R2 under different pH conditions. (**B**) Residual activity of G4 under different pH conditions. (**C**) Residual activity of G4 incubated at 65 °C for different times. (**D**) Residual activity of R2 incubated at 65 °C for different times.

Figure 6 indicated that CLEAs had only 20% activity after six cycles of utilization. However, MCLEAs retained above 70% activity after 10 cycles of utilization. For MCLEAs G4, residual activity reached 83% for 10 consecutive batches, indicating good reusability. The above results indicate that MCLEAs are more conducive to improving the stability of the enzyme.

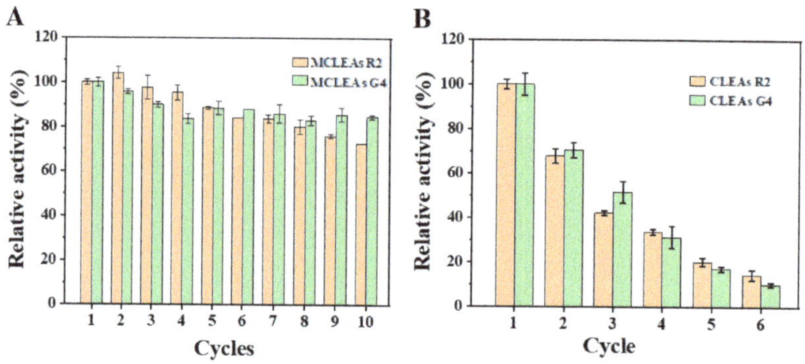

Figure 6. Reusability of MCLEAs (**A**) and CLEAs (**B**) in buffer.

2.5. Application in Icaritin Biosynthesis

As shown in Figure 7, 0.2U G4 can completely hydrolyze 1mg/mL of epimedin C within 30 min. Compared with free enzymes, MCLEAs G4 are significantly improved and can reach 129.71% of free enzymes, while the activity of CLEAs (98.27%) is basically the same as that of free enzymes. At 20 min, the conversion rate of G4 reached 80%, while MCLEAs G4 were already close to saturation. For R2, the enzyme activity was not significantly increased after fixation but decreased to a certain extent. In the optimal preparation conditions, enzyme activity recovery rates of CLEAs R2 and MCLEAs R2 were 62.93% and 92.56%, respectively. It may be owing to the relatively small pore size of rhamnosidase during immobilization, which may lead to the difficult access of macromolecular substrates and limited diffusion, resulting in decreased catalytic efficiency [41].

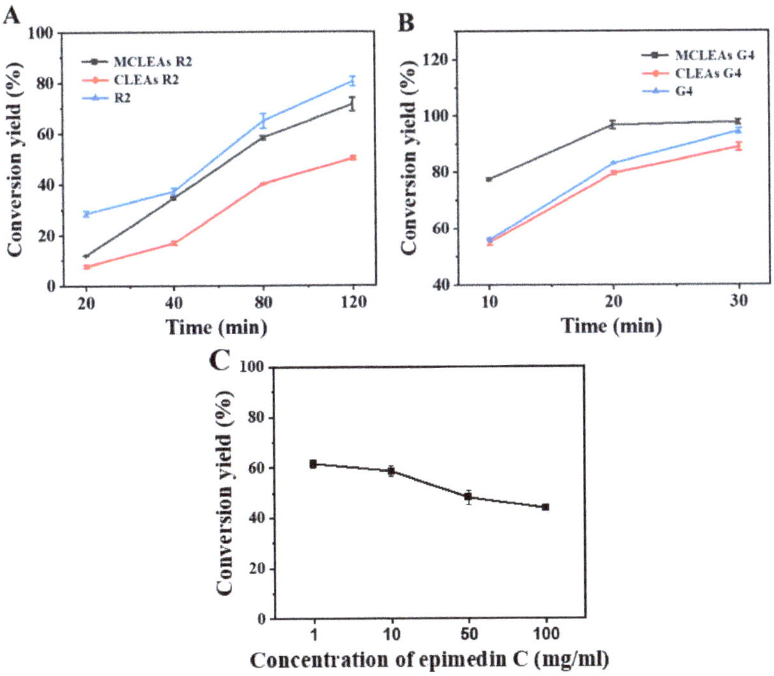

Figure 7. Conversion of epimedin C with free enzymes, CLEAs, and MCLEAs (**A**,**B**). Conversion of various concentrations of epimedin C (**C**).

The one-pot biosynthesis of icaritin was achieved by adding MCLEAs R2 and MCLEAs G4 to the optimal reaction system in a ratio of 2:1 based on the molar ratio of glucose residues and rhamnose residues present in icaritin, resulting in the conversion rate of epimedin C at 61.59%. This approach exhibits significant advantages compared to those reported in the other literature. Wang et al. utilized crude glycosidase derived from Aspergillus, strain y848, for the synthesis of icaritin from icariin, achieving a weight yield of 50.4% under optimized conditions [60]. A whole-cell catalytic experiment demonstrated that Escherichia coli, strain WPE4, exhibited an efficiency of 46.4% in converting kaempferol to icaritin [61]. Xie et al., after incubating a reaction mixture containing 1 g/L epimedin C and TpeRha at a temperature of 90 °C for 100 min, introduced Tpebgl1 into the mixture and further incubated it for an additional 50 min. Ultimately, with optimal conditions, the conversion rate of 1 g/L epimedin C reached 0.4337 g/L [9]. The efficiency of icaritin production by MCLEAs at different concentrations of epimedin C is illustrated in Figure 7C. There was no apparent substrate inhibition observed with the increasing substrate concentration.

3. Materials and Methods

3.1. Chemicals and Reagents

Epimedin C was obtained through Shanghai Aladdin Biochemical Technology Co., Ltd., Shanghai, China. p-nitrophenyl-β-D-glucopyranoside (PNPG) and p-nitrophenyl-α-rhamnopyranoside (PNPR) were obtained though Beijing Solarbio Science&Technology Co., Ltd., Beijing, China. Glutaraldehyde (GA), $FeCl_2·4H_2O$, and $FeCl_3·6H_2O$ were obtained from Macklin. Ammonium sulfate was purchased from Beijing Chemical Works (Beijing, China). The reagents employed in the subsequent experiment were of analytical purity.

3.2. Expression of Rhamnosidase and Glucosidase

In *Pichia pastoris*, rhamnosidase (R2) and glucosidase (G4) were overexpressed. pPIC9K-R2 and pPIC9K-G4, recombinant plasmids, were electroporated into *P. pastoris* GS115, followed by cultivation of the resulting recombinant cells in buffered glycerol-complex medium at 30 °C for 24 h. Enzyme expression was carried out in buffered methanol-complex medium at 30 °C by providing a daily addition of 1% methanol for 5 days, ultimately leading to the collection of R2 and G4 through centrifugation at 6000 rpm for 10 min. The enzymes were obtained from the strains cultivated and stored in our laboratory in previous work.

3.3. Enzyme Activity Assay

The enzymatic activity of R2 and G4 was assessed using 1 mg/mL PNPG or pNPR as substrate in 50 mM buffer (pH 5.0) at 55 °C. The reaction system was pre-incubated at the target temperature before the enzyme was added to initiate the timing. The reaction was conducted for 30 s and terminated by adding an equal volume of sodium carbonate solution. The absorbance of the released p-nitrophenol (pNP) was measured at 405 nm. One unit (U) of enzyme activity was defined as the amount of enzyme required to release 1 μmol of pNP per minute.

The concept of relative activity was introduced to provide a more accessible approach for assessing the catalytic activity of enzymes. It was quantified using the following equation: relative activity = A_0/A_X (where A_0 represents the highest activity within a specific group, and A_x denotes the activities of other groups).

3.4. Preparing Carboxyl-Functionalized Magnetic Fe_3O_4 Nanoparticles

On this basis, some modifications were made to prepare carboxyl-functionalized magnetic Fe_3O_4 nanoparticles [62]. We dissolve 1.195 g of $FeCl_2·4H_2O$ and 3.25 g of $FeCl_3·6H_2O$ in 50 mL of deionized water, which was transferred into a three-necked, round-bottomed flask with a volume of 100 mL, heating to 50 °C with vigorous agitation for a duration of 30 min. When the temperature was elevated to 75 °C, 6.25 mL of $NH_3·H_2O$ was added into the system and mixed for an hour. When the system temperature was raised to 85 °C, the system was supplemented with 6.25 mL (1.5 mol/L) trisodium citrate. The reaction continued for 90 min with continuous stirring until the completion of the preparation. Magnetic nanoparticles were magnetically immobilized and subsequently subjected to multiple washes with saturated sodium chloride, deionized water, and pure ethanol. Ultimately, magnetic nanoparticles were stored in 20% ethanol solution at low temperature until use.

3.5. Preparation of CLEAs and MCLEAs

CLEAs are prepared in two steps. Beforehand, the collected fermentation broth was concentrated with an ultrafiltration centrifuge tube (Solarbio Science, Beijing, China, 50 mL, and 30 kDa), and its unit activity was determined for subsequent experiments. In the first step, protein was precipitated by the addition of saturated ammonium sulfate solution in different volume ratios to form enzyme aggregates. In detail, the crude enzyme solution containing 0.25U remained at 4 °C for 2–5 h with saturated ammonium sulfate solutions in different ratios. In the subsequent phase, different glutaraldehyde concentrations were added to the formed enzyme aggregates to gain water-insoluble cross-linked enzyme aggregates. The CLEAs R2 and CLEAs G4 were finally obtained by adding glutaraldehyde to an ultimate concentration from 0.2% to 3% and cross-linking for 3 h at 16 °C.

The MCLEAs were prepared with Fe_3O_4 using the same strategy of CLEAs, but one can just use fermentation broth instead of ultrafiltrate. A total of 200 μL of fermentation broth (0.4 U/mL) and 4 mg of Fe_3O_4 were mixed with different volume ratios of saturated $(NH_4)_2SO_4$ solution; mixture was incubated at a temperature of 4 °C for a duration of 2–5 h. Eventually, different concentrations of GA were utilized to cross-link Fe_3O_4 and enzymes for a duration of 3 h at a temperature of 16 °C. The resulting precipitate was separated by magnet, then washed three times with phosphate-buffered solution (pH 5, 20 mM) to eliminate unfixed glycosidases, and stored at 4 °C.

3.6. *Stability Evaluation of Free Enzymes and Immobilized Enzymes*

To assess stability of pH and thermal stability, phosphate buffers with pH gradients from 3 to 7 and temperature gradients from 40 °C to 75 °C were prepared. Immobilized enzymes and free enzymes were incubated for 2 h in this phosphate buffer described above, and the residual activities of both the free enzymes and immobilized enzymes were assessed. Reusability of MCLEAs and CLEAs was investigated by determining the relative activity following hydrolysis of epimedin C. After every determination, MCLEAs and CLEAs were recovered through magnet or centrifugation and washed with 20 mmol/L phosphate buffer at pH 5.0.

3.7. *Characterizations*

The scanning electron microscopy (SEM, Hitachi S-4700, Hitachi, Tokyo, Japan) was utilized to document the morphologies and dimensions of CLEAs, and observation of MCLEAs was conducted using transmission electron microscopy (TEM, JEOL-JEM 2100F, JEOL-JEM 2100F) on samples that were coated with a layer of gold via sputter-coating. The magnetic properties of the samples were determined though vibrating sample magnetometer (VSM, Lake Shore 8600, Lake Shore Cryotronics, Westerville, OH, USA) under a magnetic field ranging from −15 kOe to 15 kOe at 25 °C. Participating in the immobilization process of free enzymes, CLEAs, MCLEAs, and MNPs were characterized by Fourier transform infrared (FT-IR, Shimadzu IRAffinity-1, Shimadzu Corporation, Kyoto, Japan) in the gradients of 4000–400 cm^{-1}. The identification of phases was conducted using X-ray diffraction analysis (XRD, Shimadzu XRD-6000, Shimadzu Corporation, Kyoto, Japan). The diffracting angles 2θ were scanned from 3° to 90° with a step size of 10° and a measurement time of 10 min per step.

4. Conclusions

This study systematically evaluated two immobilization techniques, cross-linked enzyme aggregates (CLEAs) and magnetic cross-linked enzyme aggregates (MCLEAs), for the enzymatic hydrolysis of epimedin C. The MCLEA system demonstrated a superior catalytic performance relative to CLEAs. Specifically, the activity of MCLEAs G4 reached 129.71%, representing a 1.4-fold increase over that of CLEAs G4 (98.27%). Similarly, the activity of R2 increased from 62.53% in CLEAs to 92.56% in MCLEAs, corresponding to a 1.56-fold enhancement. Under the optimized MCLEAs conditions, a hydrolysis conversion rate of 61.59% for epimedin C was achieved, leading to a markedly improved yield of the target product icaritin. Moreover, the immobilized enzymes exhibited enhanced operational stability and reusability, along with facile separation from the reaction system. These findings support MCLEAs as an efficient and practical strategy for the biocatalytic production of high-value natural products such as icariin.

Supplementary Materials: The following supporting information can be downloaded at: https://www.mdpi.com/article/10.3390/catal15111034/s1, Figure S1: SEM photos of CLEAs; Figure S2: FT-IR spectra of free enzyme and CLEAs; Figure S3: TEM photos of MNPs(A), MCLEAs G4(B) and MCLEAs R2(C); Figure S4: FT-IR spectra of free enzyme and MCLEAs; Figure S5: XRD patterns of MNPs and MCLEAs; Figure S6: XRD patterns of MNPs and MCLEAs.

Author Contributions: Y.Z.: Methodology, Investigation, and Writing—original draft. W.Z.: Investigation. Y.L.: Supervision, Writing—review and editing. W.F.: Investigation. H.L.: Methodology, Supervision, and Writing—review and editing. All authors have read and agreed to the published version of the manuscript.

Funding: This research was funded by the National Key Research and Development Program of China (2021YFC2102800), the National Natural Science Foundation of China (22078014), the Major Science and Technology Projects of Beijing Polytechnic (2024X005-KXD), the Supported by Beijing Natural Science Foundation (NO. 7254488).

Data Availability Statement: The original contributions presented in this study are included in the article/Supplementary Materials. Further inquiries can be directed to the corresponding author(s).

Conflicts of Interest: The authors declare that there are no conflicts of interest.

References

1. Chen, L.; Wei, X.; Gu, D.; Xu, Y.; Zhou, H. Human Liver Cancer Organoids: Biological Applications, Current Challenges, and Prospects in Hepatoma Therapy. *Cancer Lett.* **2023**, *555*, 216048. [CrossRef] [PubMed]
2. Qin, S.-Y.; Zhang, A.-Q.; Cheng, S.-X.; Rong, L.; Zhang, X.-Z. Drug Self-Delivery Systems for Cancer Therapy. *Biomaterials* **2017**, *112*, 234–247. [CrossRef] [PubMed]
3. Ion, D.; Niculescu, A.-G.; Păduraru, D.N.; Andronic, O.; Mușat, F.; Grumezescu, A.M.; Bolocan, A. An Up-to-Date Review of Natural Nanoparticles for Cancer Management. *Pharmaceutics* **2022**, *14*, 18. [CrossRef]
4. Bi, Z.; Zhang, W.; Yan, X. Anti-Inflammatory and Immunoregulatory Effects of Icariin and Icaritin. *Biomed. Pharmacother.* **2022**, *151*, 113180. [CrossRef]
5. Huong, N.T.; Son, N.T. Icaritin: A Phytomolecule with Enormous Pharmacological Values. *Phytochemistry* **2023**, *213*, 113772. [CrossRef] [PubMed]
6. Zhang, X.; Tang, B.; Wen, S.; Wang, Y.; Pan, C.; Qu, L.; Yin, Y.; Wei, Y. Advancements in the Biotransformation and Biosynthesis of the Primary Active Flavonoids Derived from Epimedium. *Molecules* **2023**, *28*, 7173. [CrossRef]
7. Liu, D.-F.; Li, Y.-P.; Ou, T.-M.; Huang, S.-L.; Gu, L.-Q.; Huang, M.; Huang, Z.-S. Synthesis and Antimultidrug Resistance Evaluation of Icariin and Its Derivatives. *Bioorg. Med. Chem. Lett.* **2009**, *19*, 4237–4240. [CrossRef]
8. Liu, F.; Wei, B.; Cheng, L.; Zhao, Y.; Liu, X.; Yuan, Q.; Liang, H. Co-Immobilizing Two Glycosidases Based on Cross-Linked Enzyme Aggregates to Enhance Enzymatic Properties for Achieving High Titer Icaritin Biosynthesis. *J. Agric. Food Chem.* **2022**, *70*, 11631–11642. [CrossRef]
9. Xie, J.; Zhang, S.; Tong, X.; Wu, T.; Pei, J.; Zhao, L. Biochemical Characterization of a Novel Hyperthermophilic α-l-Rhamnosidase from *Thermotoga petrophila* and Its Application in Production of Icaritin from Epimedin C with a Thermostable β-Glucosidase. *Process Biochem.* **2020**, *93*, 115–124. [CrossRef]
10. Dong, Y.; Zhang, S.; Lu, C.; Xu, J.; Pei, J.; Zhao, L. Immobilization of Thermostable β-Glucosidase and α-l-Rhamnosidase from Dictyoglomus Thermophilum DSM3960 and Their Cooperated Biotransformation of Total Flavonoids Extract from Epimedium into Icaritin. *Catal. Lett.* **2021**, *151*, 2950–2963. [CrossRef]
11. Liu, C.; Li, R.; Peng, J.; Qu, D.; Huang, M.; Chen, Y. Enhanced Hydrolysis and Antitumor Efficacy of Epimedium Flavonoids Mediated by Immobilized Snailase on Silica. *Process Biochem.* **2019**, *86*, 80–88. [CrossRef]
12. Shen, Y.; Wang, H.; Lu, Y.; Xu, L.; Yin, H.; Tam, J.P.; Yang, H.; Jia, X. Construction of a Novel Catalysis System for Clean and Efficient Preparation of Baohuoside I from Icariin Based on Biphase Enzymatic Hydrolysis. *J. Clean. Prod.* **2018**, *170*, 727–734. [CrossRef]

13. Guo, Y.; Wang, X.; Gao, J. Simultaneous Preparation and Comparison of the Osteogenic Effects of Epimedins A–C and Icariin from Epimedium Brevicornu. *Chem. Biodivers.* **2018**, *15*, e1700578. [CrossRef] [PubMed]
14. Martins, S.L.; Albuquerque, B.F.; Nunes, M.A.P.; Ribeiro, M.H.L. Exploring Magnetic and Imprinted Cross-Linked Enzyme Aggregates of Rhamnopyranosidase in Microbioreactors. *Bioresour. Technol.* **2018**, *249*, 704–712. [CrossRef]
15. Liu, Z.; Smith, S.R. Cross-Linked Enzyme Aggregate (CLEA) Preparation from Waste Activated Sludge. *Microorganisms* **2023**, *11*, 1902. [CrossRef]
16. Wang, S.; Zheng, D.; Yin, L.; Wang, F. Preparation, Activity and Structure of Cross-Linked Enzyme Aggregates (CLEAs) with Nanoparticle. *Enzym. Microb. Technol.* **2017**, *107*, 22–31. [CrossRef]
17. Xu, M.; Ji, D.; Deng, Y.; Agyei, D. Preparation and Assessment of Cross-Linked Enzyme Aggregates (CLEAs) of β-Galactosidase from Lactobacillus Leichmannii 313. *Food Bioprod. Process.* **2020**, *124*, 82–96. [CrossRef]
18. Shaarani, S.M.; Jahim, J.M.; Rahman, R.A.; Idris, A.; Murad, A.M.A.; Illias, R.M. Silanized Maghemite for Cross-Linked Enzyme Aggregates of Recombinant Xylanase from Trichoderma Reesei. *J. Mol. Catal. B Enzym.* **2016**, *133*, 65–76. [CrossRef]
19. Talekar, S.; Ghodake, V.; Ghotage, T.; Rathod, P.; Deshmukh, P.; Nadar, S.; Mulla, M.; Ladole, M. Novel Magnetic Cross-Linked Enzyme Aggregates (Magnetic CLEAs) of Alpha Amylase. *Bioresour. Technol.* **2012**, *123*, 542–547. [CrossRef]
20. Venkataraman, S.; Vaidyanathan, V.K. Synthesis of Magnetically Recyclable Porous Cross-Linked Aggregates of Tramates Versicolor MTCC 138 Laccase for the Efficient Removal of Pentachlorophenol from Aqueous Solution. *Environ. Res.* **2023**, *229*, 115899. [CrossRef]
21. Lucena, G.N.; dos Santos, C.C.; Pinto, G.C.; Piazza, R.D.; Guedes, W.N.; Jafelicci Júnior, M.; de Paula, A.V.; Marques, R.F.C. Synthesis and Characterization of Magnetic Cross-Linked Enzyme Aggregate and Its Evaluation of the Alternating Magnetic Field (AMF) Effects in the Catalytic Activity. *J. Magn. Magn. Mater.* **2020**, *516*, 167326. [CrossRef]
22. Escalante Morales, L.K.; Sengar, P.; Dorado Baeza, A.; Vazquez-Duhalt, R.; Chauhan, K. Enhanced Laccase Activity and Stability as Crosslinked Enzyme Aggregates on Magnetic Copper Ferrite Nanoparticles for Biotechnological Processes. *ChemCatChem* **2023**, *15*, e202301071. [CrossRef]
23. Abdulhamid, M.B.; Hero, J.S.; Zamora, M.; Gómez, M.I.; Navarro, M.C.; Romero, C.M. Effect of the Biological Functionalization of Nanoparticles on Magnetic CLEA Preparation. *Int. J. Biol. Macromol.* **2021**, *191*, 689–698. [CrossRef]
24. He, G.; Liu, H.; Yang, C.; Hu, K.; Zhai, X.; Fang, B.; Liu, K.; Zulekha; Li, D. A Comparison of Dual-Enzyme Immobilization by Magnetic Nanoparticles and Magnetic Enzyme Aggregates for Cascade Enzyme Reactions. *Biochem. Eng. J.* **2024**, *204*, 109207. [CrossRef]
25. Bounegru, A.V.; Apetrei, C. Tyrosinase Immobilization Strategies for the Development of Electrochemical Biosensors—A Review. *Nanomaterials* **2023**, *13*, 760. [CrossRef] [PubMed]
26. Ouyang, J.; Pu, S.; Wang, J.; Deng, Y.; Yang, C.; Naseer, S.; Li, D. Enzymatic Hydrolysate of Geniposide Directly Acts as Cross-Linking Agent for Enzyme Immobilization. *Process Biochem.* **2020**, *99*, 187–195. [CrossRef]
27. Bouguerra, O.M.; Wahab, R.A.; Huyop, F.; Al-Fakih, A.M.; Mahmood, W.M.A.W.; Mahat, N.A.; Sabullah, M.K. An Overview of Crosslinked Enzyme Aggregates: Concept of Development and Trends of Applications. *Appl. Biochem. Biotechnol.* **2024**, *196*, 5711–5739. [CrossRef] [PubMed]
28. Sheldon, R. CLEAs, Combi-CLEAs and 'Smart' Magnetic CLEAs: Biocatalysis in a Bio-Based Economy. *Catalysts* **2019**, *9*, 261. [CrossRef]
29. Abdul Wahab, M.K.H.; El-Enshasy, H.A.; Bakar, F.D.A.; Murad, A.M.A.; Jahim, J.M.; Illias, R.M. Improvement of Cross-Linking and Stability on Cross-Linked Enzyme Aggregate (CLEA)-Xylanase by Protein Surface Engineering. *Process Biochem.* **2019**, *86*, 40–49. [CrossRef]
30. Barbosa, O.; Ortiz, C.; Berenguer-Murcia, Á.; Torres, R.; Rodrigues, R.C.; Fernandez-Lafuente, R. Glutaraldehyde in Bio-Catalysts Design: A Useful Crosslinker and a Versatile Tool in Enzyme Immobilization. *RSC Adv.* **2013**, *4*, 1583–1600. [CrossRef]
31. Zhen, Q.; Wang, M.; Qi, W.; Su, R.; He, Z. Preparation of β-Mannanase CLEAs Using Macromolecular Cross-Linkers. *Catal. Sci. Technol.* **2013**, *3*, 1937. [CrossRef]
32. Panwar, D.; Kaira, G.S.; Kapoor, M. Cross-Linked Enzyme Aggregates (CLEAs) and Magnetic Nanocomposite Grafted CLEAs of GH26 Endo-β-1,4-Mannanase: Improved Activity, Stability and Reusability. *Int. J. Biol. Macromol.* **2017**, *105*, 1289–1299. [CrossRef]
33. Nadar, S.S.; Rathod, V.K. Magnetic Macromolecular Cross Linked Enzyme Aggregates (CLEAs) of Glucoamylase. *Enzym. Microb. Technol.* **2016**, *83*, 78–87. [CrossRef]
34. Schoevaart, R.; Wolbers, M.W.; Golubovic, M.; Ottens, M.; Kieboom, A.P.G.; van Rantwijk, F.; van der Wielen, L.A.M.; Sheldon, R.A. Preparation, Optimization, and Structures of Cross-Linked Enzyme Aggregates (CLEAs). *Biotechnol. Bioeng.* **2004**, *87*, 754–762. [CrossRef]
35. Li, L.; Li, G.; Cao, L.; Ren, G.; Kong, W.; Wang, S.; Guo, G.; Liu, Y.-H. Characterization of the Cross-Linked Enzyme Aggregates of a Novel β-Galactosidase, a Potential Catalyst for the Synthesis of Galacto-Oligosaccharides. *J. Agric. Food Chem.* **2015**, *63*, 894–901. [CrossRef]

36. Rathod, A.G.; Tiwari, P.; Shaily, J.; Tiwari, S. Stabilizing Effect of Quercetin upon Bovine Serum Albumin as a Model Protein. *Colloids Surf. B Biointerfaces* **2025**, *252*, 114663. [CrossRef]
37. Fazary, A.E.; Ismadji, S.; Ju, Y.-H. Biochemical Studies on Native and Cross-Linked Aggregates of Aspergillus Awamori Feruloyl Esterase. *Int. J. Biol. Macromol.* **2009**, *44*, 240–248. [CrossRef]
38. Khatik, A.G.; Jain, A.K.; Muley, A.B. Preparation, Characterization and Stability of Cross Linked Nitrilase Aggregates (Nitrilase-CLEAs) for Hydroxylation of 2-Chloroisonicotinonitrile to 2-Chloroisonicotinic Acid. *Bioprocess Biosyst. Eng.* **2022**, *45*, 1559–1579. [CrossRef] [PubMed]
39. Joseph, J.E.; Mary, P.R.; Haritha, K.V.; Panwar, D.; Kapoor, M. Soluble and Cross-Linked Aggregated Forms of α-Galactosidase from Vigna Mungo Immobilized on Magnetic Nanocomposites: Improved Stability and Reusability. *Appl. Biochem. Biotechnol.* **2021**, *193*, 238–256. [CrossRef]
40. Wei, B.; Liu, F.; Liu, X.; Cheng, L.; Yuan, Q.; Gao, H.; Liang, H. Enhancing Stability and By-Product Tolerance of β-Glucuronidase Based on Magnetic Cross-Linked Enzyme Aggregates. *Colloids Surf. B Biointerfaces* **2022**, *210*, 112241. [CrossRef]
41. Ifko, D.; Vasić, K.; Knez, Ž.; Leitgeb, M. (Magnetic) Cross-Linked Enzyme Aggregates of Cellulase from T. Reesei: A Stable and Efficient Biocatalyst. *Molecules* **2023**, *28*, 1305. [CrossRef]
42. Li, K.; Wang, J.; He, Y.; Cui, G.; Abdulrazaq, M.A.; Yan, Y. Enhancing Enzyme Activity and Enantioselectivity of Burkholderia Cepacia Lipase via Immobilization on Melamine-Glutaraldehyde Dendrimer Modified Magnetic Nanoparticles. *Chem. Eng. J.* **2018**, *351*, 258–268. [CrossRef]
43. Bian, H.; Cao, M.; Wen, H.; Tan, Z.; Jia, S.; Cui, J. Biodegradation of Polyvinyl Alcohol Using Cross-Linked Enzyme Aggregates of Degrading Enzymes from Bacillus Niacini. *Int. J. Biol. Macromol.* **2019**, *124*, 10–16. [CrossRef] [PubMed]
44. Migneault, I.; Dartiguenave, C.; Bertrand, M.J.; Waldron, K.C. Glutaraldehyde: Behavior in Aqueous Solution, Reaction with Proteins, and Application to Enzyme Crosslinking. *Biotechniques* **2004**, *37*, 790–802. [CrossRef]
45. Wang, H.; Han, S.; Wang, J.; Yu, S.; Li, X.; Lu, L. Preparation and Synthetic Dye Decolorization Ability of Magnetic Cross-Linked Enzyme Aggregates of Laccase from Bacillus Amyloliquefaciens. *Bioprocess Biosyst. Eng.* **2021**, *44*, 727–735. [CrossRef]
46. Lucena, G.N.; dos Santos, C.C.; Pinto, G.C.; da Rocha, C.O.; Brandt, J.V.; de Paula, A.V.; Jafelicci Júnior, M.; Marques, R.F.C. Magnetic Cross-Linked Enzyme Aggregates (MCLEAs) Applied to Biomass Conversion. *J. Solid State Chem.* **2019**, *270*, 58–70. [CrossRef]
47. Talekar, S.; Joshi, A.; Kambale, S.; Jadhav, S.; Nadar, S.; Ladole, M. A Tri-Enzyme Magnetic Nanobiocatalyst with One Pot Starch Hydrolytic Activity. *Chem. Eng. J.* **2017**, *325*, 80–90. [CrossRef]
48. Patil, S.S.; Rathod, V.K. Combined Effect of Enzyme Co-Immobilized Magnetic Nanoparticles (MNPs) and Ultrasound for Effective Extraction and Purification of Curcuminoids from Curcuma Longa. *Ind. Crops Prod.* **2022**, *177*, 114385. [CrossRef]
49. Sahu, A.; Badhe, P.S.; Adivarekar, R.; Ladole, M.R.; Pandit, A.B. Synthesis of Glycinamides Using Protease Immobilized Magnetic Nanoparticles. *Biotechnol. Rep.* **2016**, *12*, 13–25. [CrossRef]
50. Singh, V.; Rakshit, K.; Rathee, S.; Angmo, S.; Kaushal, S.; Garg, P.; Chung, J.H.; Sandhir, R.; Sangwan, R.S.; Singhal, N. Metallic/Bimetallic Magnetic Nanoparticle Functionalization for Immobilization of α-Amylase for Enhanced Reusability in Bio-Catalytic Processes. *Bioresour. Technol.* **2016**, *214*, 528–533. [CrossRef]
51. Sandler, S.E.; Fellows, B.; Mefford, O.T. Best Practices for Characterization of Magnetic Nanoparticles for Biomedical Applications. *Anal. Chem.* **2019**, *91*, 14159–14169. [CrossRef]
52. Chi, M.-C.; Huang, Y.-F.; Lu, B.-Y.; Lin, M.-G.; Wang, T.-F.; Lin, L.-L. Magnetic Cross-Linked Enzyme Aggregates of a Transpeptidase-Specialized Variant (N450D) of Bacillus Licheniformis γ-Glutamyl Transpeptidase: An Efficient and Stable Biocatalyst for l-Theanine Synthesis. *Catalysts* **2021**, *11*, 243. [CrossRef]
53. Zhang, K.; Yang, W.; Liu, Y.; Zhang, K.; Chen, Y.; Yin, X. Laccase Immobilized on Chitosan-Coated Fe3O4 Nanoparticles as Reusable Biocatalyst for Degradation of Chlorophenol. *J. Mol. Struct.* **2020**, *1220*, 128769. [CrossRef]
54. Guzik, U.; Hupert-Kocurek, K.; Wojcieszyńska, D. Immobilization as a Strategy for Improving Enzyme Properties-Application to Oxidoreductases. *Molecules* **2014**, *19*, 8995–9018. [CrossRef] [PubMed]
55. Nawawi, N.N.; Hashim, Z.; Manas, N.H.A.; Azelee, N.I.W.; Illias, R.M. A Porous-Cross Linked Enzyme Aggregates of Maltogenic Amylase from Bacillus Lehensis G1: Robust Biocatalyst with Improved Stability and Substrate Diffusion. *Int. J. Biol. Macromol.* **2020**, *148*, 1222–1231. [CrossRef]
56. Moradi, S.; Khodaiyan, F.; Hadi Razavi, S. Green Construction of Recyclable Amino-Tannic Acid Modified Magnetic Nanoparticles: Application for β-Glucosidase Immobilization. *Int. J. Biol. Macromol.* **2020**, *154*, 1366–1374. [CrossRef] [PubMed]
57. Murphy, J.; Ryan, M.P.; Walsh, G. Purification and Characterization of a Novel β-Galactosidase From the Thermoacidophile Alicyclobacillus Vulcanalis. *Appl. Biochem. Biotechnol.* **2020**, *191*, 1190–1206. [CrossRef]
58. Cheng, G.; Pi, Z.; Zheng, Z.; Liu, S.; Liu, Z.; Song, F. Magnetic Nanoparticles-Based Lactate Dehydrogenase Microreactor as a Drug Discovery Tool for Rapid Screening Inhibitors from Natural Products. *Talanta* **2020**, *209*, 120554. [CrossRef]
59. Salgın, S.; Salgın, U. Bioconjugation of Magnetic Iron Oxide Nanoparticles and Cross-Linked Enzyme Aggregates: Characterization and Comparison of CLEAs and Magnetic CLEAs. *Curr. Nanosci.* **2017**, *13*, 149–158. [CrossRef]

60. Wang, Z.; Liu, C.; Yu, H.; Wu, B.; Huai, B.; Zhuang, Z.; Sun, C.; Xu, L.; Jin, F. Icaritin Preparation from Icariin by a Special Epimedium Flavonoid-Glycosidase from *Aspergillus* Sp.Y848 Strain. *J. Microbiol. Biotechnol.* **2022**, *32*, 437–446. [CrossRef]
61. Wang, P.; Li, C.; Li, X.; Huang, W.; Wang, Y.; Wang, J.; Zhang, Y.; Yang, X.; Yan, X.; Wang, Y.; et al. Complete Biosynthesis of the Potential Medicine Icaritin by Engineered *Saccharomyces cerevisiae* and *Escherichia coli*. *Sci. Bull.* **2021**, *66*, 1906–1916. [CrossRef] [PubMed]
62. He, J.; Sun, S.; Zhou, Z.; Yuan, Q.; Liu, Y.; Liang, H. Thermostable Enzyme-Immobilized Magnetic Responsive Ni-Based Metal–Organic Framework Nanorods as Recyclable Biocatalysts for Efficient Biosynthesis of S-Adenosylmethionine. *Dalton Trans.* **2019**, *48*, 2077–2085. [CrossRef] [PubMed]

MDPI AG
Grosspeteranlage 5
4052 Basel
Switzerland
Tel.: +41 61 683 77 34

Catalysts Editorial Office
E-mail: catalysts@mdpi.com
www.mdpi.com/journal/catalysts

www.ingramcontent.com/pod-product-compliance
Lightning Source LLC
Chambersburg PA
CBHW062052220926
51850CB00010B/2745
9783725869121